PROTOPLASMATOLOGIA
HANDBUCH
DER PROTOPLASMAFORSCHUNG

HERAUSGEGEBEN VON

L. V. HEILBRUNN UND F. WEBER
PHILADELPHIA   GRAZ

MITHERAUSGEBER

W. H. ARISZ - GRONINGEN · H. BAUER - WILHELMSHAVEN · J. BRACHET - BRUXELLES · H. G. CALLAN - ST. ANDREWS · R. COLLANDER - HELSINKI · K. DAN - TOKYO · E. FAURÉ - FREMIET - PARIS · A. FREY-WYSSLING - ZÜRICH · L. GEITLER - WIEN · K. HÖFLER - WIEN · M. H. JACOBS - PHILADELPHIA · D. MAZIA - BERKELEY · A. MONROY - PALERMO · J. RUNNSTRÖM - STOCKHOLM · W. J. SCHMIDT - GIESSEN · S. STRUGGER - MÜNSTER

BAND X

PATHOLOGIE DES PROTOPLASMAS

5 a

MORPHOLOGY AND PHYSIOLOGY OF PLANT TUMORS

WIEN

SPRINGER-VERLAG

1958

MORPHOLOGY AND PHYSIOLOGY OF PLANT TUMORS

BY

ARMIN C. BRAUN AND TOM STONIER

NEW YORK

WITH 7 FIGURES

WIEN

SPRINGER-VERLAG

1958

ISBN-1-13: 978-3-211-80492-6   e-ISBN-13978-3-7091-5536-3

DOI: 10.1007/978-3-7091-5536-3

Morphology and Physiology of Plant Tumors

By

ARMIN C. BRAUN and TOM STONIER

The Rockefeller Institute for Medical Research, New York, N.Y.

With 7 Figures

Contents

Introduction

The tumor problem is basically a problem of growth. It is, moreover, a problem of the uncontrolled growth of altered cells by more or less random division. Rate of growth does not in itself characterize the tumor cell since certain normal cell types may grow and divide at considerably faster rates than do most tumor cells. The tumor cell need not, however, be lacking the capacity to differentiate and functionate. Certain malignant teratomata may retain these properties to a striking degree. The one biological characteristic that distinguishes the tumor cell from the normal cell is its capacity for the uncontrolled or autonomous growth within an organism. Without this there would be no tumors.

Autonomy exhibited by neoplastic growth has many gradations. At one extreme are found the benign tumors which grow slowly and remain localized in the host; at the other are the most malignant cancers that invade neighboring tissues and spread throughout an organism by metastasis. The nature of autonomy remains a foremost problem in experimental oncology.

The fundamental similarity of cells and cellular processes is commonly recognized. Since the capacity for autonomous growth is potentially a property of the cell itself, it is likely to be expressed in the cells of all higher organisms whether they are of animal or plant origin. It does not necessarily follow, however, that the cellular mechanism in terms of specific chemical reactions is identical in members of the two kingdoms. It is, nevertheless, probable that advances in one field will assist in clarifying problems in related fields of study.

Plants have for many years provided interesting and unique experimental material for studies of both normal and abnormal growth and development, and as such have become recognized as objects of fundamental value for researches dealing with the tumor problem generally. As early as 1910 C. O. Jensen, the father of modern experimental cancer research, pointed out that the formation of crown gall tumors in beets appeared to be based upon a continuous, abnormal, proliferative capacity of certain cells. Furthermore, the effect upon the growth of the beet, the abnormal chemical relationship of the tumor to the normal tissue, and the fact that the tumor was transplantable "... remind one so much of the malignant tumors of animals, that a closer study of the biological relationships of this tumor would undoubtedly be profitable" (Jensen 1910).

In comparing plant and animal tumors it must be remembered, however, that there are certain developmental and functional characteristics commonly used in the differentiation of animal cancers that are more or less restricted to animals and cannot, therefore, be carried over and applied to plant tumors. These have been dealt with in detail by WHITE and BRAUN (1942) and by BLACK (1949), and will not be considered further here. The most essential characteristic of being able to grow independently of any morphogenetic restraint, upon which all of the other diagnostic features must ultimately depend, is, however, equally capable of expression in neoplasia of all higher organisms since it is a characteristic of the cell itself.

One striking aspect of tumor genesis is the multiplicity of diverse agencies that are seemingly capable of accomplishing essentially the same end result. Radiant energy, irritation, carcinogenic chemicals, parasitic organisms, and viruses have all been shown to serve as incitors of tumors in animals. The effectiveness of these various factors in eliciting tumor formation appears to be a function of the hereditary constitution of the host. Many of these same agencies have also been found to be concerned etiologically in the inception of tumors in plants.

In the present discussion the authors will confine themselves to a consideration of three non-self-limiting neoplastic diseases of plants each of which has a different and distinct etiology. These are 1) the crown gall disease in which some as yet uncharacterized principle elaborated by a microorganism regularly converts normal plant cells to tumor cells; 2) BLACK's wound tumor disease which is of known viral origin; and 3) the KOSTOFF tumors which commonly arise as a result of irritation in certain interspecific hybrids within the genus *Nicotiana* (Fig. 1). In addition, a fourth phenomenon will be dealt with, namely that of GAUTHERET's habituation, which is encountered in tissue culture studies. There are other true neoplasms of higher plants that have not as yet been extensively studied. Most interesting among these is WHITE's spruce tumor (WHITE and MILLINGTON 1954 a, b; REINERT and WHITE 1956), the inciting cause of which is as yet unknown. Two additional overgrowths have recently been described as occurring in lower plant forms. THOMAS, EVANS and HUGHES (1956) reported the development of tumorous overgrowths on fructifications of the cultivated mushroom *Agaricus campestris* fr. va. *bisporus*, following brief exposure of the pilius to volatile mineral oils and diesel oil vapors. Similarly the slime mold *Physarum polycephalum* was reported by SETÄLÄ. LUNDBOM and HOLSTI (1957) to develop "tumors" following application of methylcholanthrene, other carcinogenic hydrocarbons, as well as urethane.

Aside from the true plant tumors there exist many other interesting types of growth abnormalities, the descriptions of which have produced a voluminous literature. Most, if not all of these, appear to be of a self-limiting nature and will therefore not be discussed here. The reader is referred instead to the following articles and books on the subject: BUTLER (1930), KÜSTER (1911), MANIL (1950), and WHITE (1951), as well as to the several articles appearing in the Brookhaven Symposia in Biology No. 6 (1954).

It is the purpose of the present review to correlate morphological and physiological observations with the view in mind of evolving certain morphogenetic concepts that may serve to help explain the development of plant tumors. In the general discussion of this article an attempt has been made to relate these findings in plant oncology to the cancer problem generally.

The Crown Gall Disease

Introduction

Crown gall is a neoplastic disease of plants that is initiated by a specific bacterium. This bacterium possesses the rather remarkable ability of being able to transform normal plant cells to tumor cells in short periods of time. Once the cellular alteration has been accomplished, the continued abnormal proliferation of the affected host cells becomes an automatic process that is completely independent of the inciting bacteria. The transformation of normal cells to tumor cells in this disease is dependent not only upon the presence of a tumor-inducing principle elaborated by the bacteria, but also upon host cells that are conditioned and thus capable of responding to the carcinogenic influence of that principle. Normal plant cells are, for the most part, susceptible to transformation for only a short period of time following the stimulus of a wound.

Bacteria-free tissues isolated from both primary and secondary crown gall tumors commonly grow profusely and indefinitely on a culture medium that does not support the continued growth of normal cells of the type from which the tumor cells were derived. Fragments of such sterile tumor tissue grafted into a healthy host develop again into characteristic crown gall tumors that are indistinguishable morphologically and cytologically from those originally induced by the bacteria. Crown gall tumor tissue is characterized by a capacity for unlimited autonomous growth both *in vitro* and *in situ*. Autonomy appears to be the result of the continued production by the tumor cell of greater than regulatory amounts of intracellular hormones concerned with growth accompanied by cell division. Growth-substance synthesis is precisely regulated in all normal plant cells.

The first description of the crown gall disease, as far as the authors are aware, was given by Hlubek in 1839, as cited by Jensen (1910)[1]. At the turn of the century the cause of the disease was a subject of considerable speculation, a variety of agents being blamed for its inception: unrecognized soil factors, a new species of slime mold, injury response, and a bacterial agent (as reviewed by Heald 1933). In 1907, Smith and Townsend (1907 a, b) reported that they had isolated the inciting pathogen, a gram-negative bacterium, from the galls of the Paris daisy (*Chrysanthemum frutescens* L.). To the isolated organism they gave the name *Bacterium tumefaciens;* this name has been revised to *Agrobacterium tumefaciens*

---

[1] No reference to Hlubek's studies, however, is given by Jensen.

(SMITH and TOWN.) CONN (BERGEY 1948) in accordance with more modern concepts of bacterial taxonomy.

The isolation and characterization of a specific bacterium as the causative agent of this neoplastic disease in 1907 attracted considerable interest among pathologists generally. It should be recalled that at the time of this discovery no animal tumor had yet been produced experimentally. The crown gall disease has drawn the attention of a large number of investigators not only because of its economic importance but more particularly because of its theoretical implications in the field of oncology. This interest can be traced to ERWIN F. SMITH who repeatedly asserted that crown gall is a plant cancer and is analogous to the human cancer (SMITH 1912 b; 1913; 1916 a, b, c, e; 1922; 1923; 1924; 1925 a, b, c; 1926). It was SMITH who pointed to the basic similarities between crown gall and cancer. Unfortunately, the issue became beclouded when SMITH stated categorically that, since crown gall was caused by bacteria, cancer in animals and humans must also be due to an infective agent. This opinion was criticized by ROBINSON (1927) when no evidence favoring such a view could be established. In fact, although many workers had originally accepted SMITH's idea that crown gall was analogous to cancer in animals (the term "plant cancer" being used freely both in Europe and America), by the late nineteen thirties the analogy was being regarded with more and more reservation. JENSEN's (1910, 1918) brilliant experiments indicating the occurrence of sterile sugar beet tumors were completely ignored. As shall be discussed below, SMITH was in error. The bacteria only initiate the tumors by transforming normal host cell into tumor cells. The continued abnormal growth of the tumor cells becomes at an early period completely independent of the inciting bacteria. Crown gall is not, as SMITH had initially postulated, simply a bacterial-stimulated hyperplasia but is a true cancerous neoplasm of plants.

The crown gall problem has been reviewed on numerous occasions. For a comprehensive survey of the subject the reader is referred to two recent review articles (BRAUN 1954 a; KLEIN and LINK 1955), and STAPP's recent (1956) book on bacterial diseases in plants. In view of the adequate general coverage of the crown gall disease, the present discussion will limit itself primarily to morphological considerations of the disease as well as to those physiological aspects of the problem that are concerned specifically with growth accompanied by cell division. Other topics are included only as background for the general discussion.

The fundamental nature of the disease is evidenced by the fact that plants in at least 142 genera, belonging to 61 widely separated botanical families, are susceptible to crown gall (ELLIOTT 1951). An apple tree is susceptible, as is a cactus, or a sunflower. Many conifers and most dicotyledonous plants respond; however, no monocot shows unequivocal responses. STAPP (1938) has reported tumors on *Asparagus Sprengeri* Regel, as have BROWN and WEISS (1937); JAKOWSKA (1949) noted responses on onion. One report suggests that certain algae respond as well ((CHEMIN

1937). Several studies have indicated that responses varying from septicemia to actual cell proliferation may be evoked in various animals, both vertebrates and invertebrates, upon inoculation with *Agrobacterium tumefaciens*. However, except for the production of inflammatory reactions or septicemia, such reports have never been substantiated.

Morphological aspects

The morphology of the crown gall tumor may vary from slight, hardly perceptible swellings to large, highly irregular, convoluted tumor masses which may be much larger than the organ supporting such a tumor. Tumors weighing in excess of one hundred pounds have been reported on certain perennial plants. A second type of manifestation involves the production of abortive and malformed supernumerary organs (teratomata).

Many excellent photographs of the various types of tumors on innumerable hosts may be found in the early crown gall literature (*e. g.* Smith, Magnus, M. Levine, and many others). Smith initially recognized two types: a hard woody, and a soft parenchymatous type, a distinction which may also be applied to tumor tissue excised and cultivated *in vitro*. De Ropp (1947 c) obtained such types from the primary tumors of sunflower. The woody type is more organized; on prolonged cultivation, however, it becomes converted to the disorganized, parenchymatous type. Sunflower stem tumors may vary in another way, that is, convoluted *vs.* smooth. Generally, an infected needle puncture produces on sunflower large, irregular, convoluted tumors which are covered by a necrotic, corky pseudocicatrix, the tumor having burst through the epidermis relatively early in its development. Occasionally, however, the tumor does not burst through but continues to expand internally, resulting in the formation of a large, smooth, globose swelling. If the death of the host does not intervene in the subsequent development of the tumor, the growth will ultimately break through the epidermis and begin to take on the appearance of the convoluted tumor.

Aside from the production of a tumor at the site of inoculation, secondary tumors (Fig. 1 *A*) are found to occur at some distance from the point of inoculation. This phenomenon is, however, confined to relatively few plant species. Smith interpreted the formation of these secondary tumors as being the result of metastases. He considered the primary tumor to invade the surrounding tissue, traverse the stem, and form new tumors at other sites. Later, Smith (1922) modified this concept to include the process of appositional growth as a factor in the spread of the neoplastic tissue. Riker (1923 b) and Robinson and Walkden (1923) presented evidence which indicated that secondary tumors and tumor strands result from inoculations close to, or into, the meristematic region of the stem, the inciting bacteria being spread as a result of the elongation of the cells in the stem apex. These authors considered all secondary tumors to result from such a process, a view which is doubtless correct in some instances. Levine (1925) was unable to observe connecting strands between primary and secondary tumors and concluded, as had Riker and Robinson and Walkden,

Fig. 1. *A*, A primary and a secondary crown gall tumor on a sunflower plant (*Helianthus annuus* L.). *B*, Tumors of the type that arise spontaneously in certain interspecific hybrids within the genus *Nicotiana* (*N. langsdorffii × N. suaveolens*). *C*, Crown gall teratoma that originated at the cut stem surface of a tobacco plant (*Nicotiana tabacum* L.). *D*, Virus-induced tumor on sweet clover (*Melilotus alba* Desr.).
(Photographs by J. A. CARLILE.)

that the two areas were inoculated at the same time but were subsequently separated by elongating non-infected tissue. Suit and Eardley (1935) suggested that many of the so-called tumor strands reported by Smith et al. in reality represent internal primary tumors formed in the protoxylem region at points where the vessels had ruptured as a result of elongation of the stem tissue.

A different mechanism is involved in the sunflower (Braun 1941). Secondary tumors may be obtained by introducing the bacteria several inches below the apical bud into internodes which have elongated to, or almost to, their full length. This precludes the possibility of secondary tumor formation as a result of elongation of immature tissues. The tumor strands and secondary tumors are invariably associated with the xylem; the strands are not direct longitudinal outgrowths of the primary tumor, but develop laterally from the xylem into the pith as if the stimulus emanated from the xylem. With some attenuated bacterial strains, no distinct primary tumor forms at all, only a thickening of the stem. On such stems, secondary tumors will occasionally form even though the characteristic primary tumor fails to develop. Secondary tumors have also been reported *in vitro* by de Ropp (1947 b, 1948 a, b) who grafted bacteria-free sunflower crown gall tissue onto normal stem fragments. Frequently, adjacent to the primary (grafted) tumor, a new "secondary" outgrowth appeared which de Ropp believes to originate from the normal stem fragment. On one occasion, a new tumor appeared at the end opposite the stem fragment to which the tumor had been grafted. The mechanism of secondary tumor formation in sunflower remains obscure.

In sunflower, the infection of a transverse puncture wound through the stem does not commonly result in tumors directly lethal to the plant. although the handicap may cause the plant to succumb more readily to other agents. Only if drastic wounding such as that caused by a longitudinal puncture from the tip down into the stem of a young plant is accompanied by infection does the resulting massive tumor growth so disorganize the plant as to cause its early death. Other hosts, however, may be much more susceptible. *Nicotiana rustica* L. will produce large tumors when the stem is infected with a highly virulent strain, by means of a transverse puncture. Plants bearing such tumors die prematurely. On the other hand, many other species of plants do not appear to be appreciably handicapped by the presence of tumors, and in some instances the tumors may actually regress. This is particularly true of plants infected with bacterial strains of only moderate or weak virulence. The susceptibility or resistance of the host is not associated with any obvious taxonomic relationship. Thus, for example, Magnus (1915) described a wide spectrum of responses to inoculation in the genus *Pelargonium* including the fact that *P. peltatum* Ait. failed to respond at all. Similarly. Jensen (1918) has reported that sugar beets produce enormous irregular convoluted tumors. while the mangel produces roundish smooth tumors which as a rule have no detrimental effects upon the growth of the plant. and in the common garden beet the tumors attain only a slight size.

Another variation in response, which may differ from species to species when semi-virulent bacterial strains are used, involves the degree of organization or disorganization exhibited by the induced growth. As early as 1915, MAGNUS reported various types of teratomata, describing such growths as those resembling witches' brooms (*Begonia*), cauliflowers and tube-leaves (*Pelargonium*), porcupines (*Pelargonium* hybrid) as well as the production of recognizable organs such as roots and shoots on *Pelargonium* and adventitious buds on potato stems. Shortly thereafter, SMITH (1916 d) reported that he, too, was able to obtain such teratomata by inoculating into the leaf axils of young growing plants whose lateral buds are in a dormant state (*Pelargonium, Nicotiana, Lycopersicum, Citrus, Ricinus*), as well as by inoculating the midvein of young tobacco leaves. The following year SMITH (1917) discussed the subject of plant teratomata (embryomas) more fully, reporting such formations to occur on 20 plant species belonging to 15 different families. Since that time many workers have reported on the formation of teratomata, including BRAUN (1948), M. LEVINE (1919, 1923 b, 1924) and KÜSTER (1926).

Attempts to analyze the factors which influence the formation of teratomata may be traced back to SMITH (1917). As already indicated, SMITH observed a position effect, that is, whether a plant responded to infection by the formation of a tumor or of a teratoma depended on where and how the susceptible plant was inoculated. For example, in some plants inoculating into the leaf axil would tend to produce a teratoma, whereas inoculating into the mid-internode would tend to produce a tumor. He also attempted to define which histological region originated the tissue by decapitating tobacco stems and inoculating specific tissue by means of a fine needle. Unfortunately, he obtained tumors at only 4 per cent of points of inoculation in these studies. The experiment nevertheless indicated that inoculations into the cambium, protoxylem, or cortex could produce teratomata. It must be remembered, however, that these experiments are inconclusive since the bacteria are known to migrate, and indeed, SMITH stated that when tumors occurred they appeared first as swellings in the deeper parts of the tissues and not at the cut (and accidentally disinfected) surface.

SMITH believed that teratomata arose when infected "conjunctive" (*i. e.* non-vascular) tissue close to totipotent or pluripotent cells stimulated such cells to grow, thereby forming a complex tumor. In contrast to this view that teratomata originate from a mixture of normal and tumor cells, is the view first expressed by LEVIN and LEVINE (1918, 1920) that the differentiated tissue arises from new growth of the tumor. LEVINE (1919, 1924) stated that in the leaves of *Bryophyllum calycinum* Salisb. the pluripotent cells are not stimulated by crown gall, on the contrary, that they are inhibited. SMITH (1921) countered by describing and presenting photographs of complex overgrowths which obviously implied that the lateral buds in the leaf axils of *Bryophyllum* (also the meristems in the notches of the leaves) had been stimulated by the crown gall inoculation. LEVINE (1923 b), working on tobacco, then classified teratomata into two types: *1*) a type, indicated by

Smith, which Levine called axillary leafy crown galls and which were
obtained upon inoculation of the leaf axil in such a manner as to elicit
a combination of tumor and aberrant axillary bud development, and 2) a
globular type originating from stem internodes which became covered
with small leaf-like structures.

Locke, Riker, and Duggar (1938) obtained adventitious shoots on the
surfaces of cut tomato stem internodes when such a wound was inoculated
with an attenuated strain. A tendency to shoot-development could also be
observed when the cut surface was inoculated with the virulent strain
provided that the axillary buds had been removed. In these instances,
however, adventitious bud development probably involved normal tissues
since such buds developed prior to the appearance of gall tissue in those
plants whose axillary buds had not been cut off and whose cut surface was
inoculated with the attenuated strain. In plants whose axillary buds had
been excised, adventitious buds also developed from the cut surface of
uninoculated controls.

An attempt to analyze more precisely some of the conditions which
determine tumor morphology led to the following conclusions (Braun 1953):
1) of importance is the bacterial strain used. Under suitable conditions,
a highly virulent strain tends to produce disorganized tumors; a moderately
virulent strain tends to produce teratomata in hosts the cells of which
possess well-developed regenerative capacities. This is not, however, strictly
a function of virulence, because some strains which are less virulent than
those known to produce teratomata will produce small but disorganized
growths under the same condititons. 2) Of equal importance is the
reactivity of the host tissue. This appears to be a function of at least two
factors: (a) the inherent potentialities of the host cell, and (b) the relative
position that the tumor occupies in a host. When, for example, a tobacco
plant is cut through an internode in the middle of the plant and both
freshly cut stem surfaces are inoculated with a moderately virulent strain
of the crown gall bacterium, typical unorganized tumors develop at the
basal end of the upper cutting. The other inoculated surface, now the
tip of the lower half of the plant, gives rise to a complex tumor or teratoma.
Cells at the two cut surfaces prior to the time of their separation possessed
the same potentialities since they were adjoining cells of the same stem.
After severing the stem, however, cells below the cut became apical cells of
the basal portion of the original plant, while cells above the cut became
basal cells of the upper cutting.

Bacteria-free tissue fragments isolated either from teratomata initiated
at the cut stem tips or from the unorganized tumors that developed at the
basal ends of the tobacco cuttings were similar in growth pattern when
cultivated *in vitro*, even though derived from morphologically very dis-
similar overgrowths. Cells isolated from either type of tumor grew
profusely and retained indefinitely in culture the capacity to organize small
abnormal leaves and buds. When fragments of such sterile teratoma
tissues were grafted to the tips of healthy tobacco plants, they developed
into typical teratomata. When, on the other hand, similar fragments were

implanted at cambial level into internodes of tobacco plants containing functional apical buds, these fragments developed into undifferentiated and unorganized tumors. Grafting experiments such as those reported above demonstrate therefore that the ability of pluripotent tobacco tumor cells to organize is determined largely by the relative position that they occupy in the host. The restraining influence that a host containing a functional apical bud exercises on the organizational capacity of these tumor tissues is lost when the tissues are grown in culture. That the suppression of organization is a reflection of the host hormone physiology is indicated by the observation (STONIER, unpublished) that if a tobacco plant is cut through an internode and then inoculated with a moderately virulent strain just below the cut surface, a complex tumor will develop. When, on the other hand, a growth substance of the auxin type is applied to the cut surface at the time of inoculation, a typical unorganized tumor is formed.

When experiments of the type reported above were carried out with a highly virulent instead of a moderately virulent strain of crown gall bacteria, typical unorganized crown gall tumors developed at both cut stem surfaces of severed tobacco plants. Bacteria-free fragments isolated from tumors that developed at the tips and from the basal ends of the tobacco cuttings grew profusely in culture but showed not the slightest tendency to organize leaves and buds. Sterile tumor tissue of this type, grafted either to the cut stem tips or at cambial level into internodes of tobacco plants containing a functional apical bud, grew as typical unorganized tumors. Results such as these indicate that, although the tumor-inducing principle elaborated by the moderately virulent strain of bacteria is incapable of suppressing the cellular factors concerned with differentiation and organization in pluripotent tobacco cells, the principle associated with the highly virulent strain completely overwhelms those factors. The teratoma tissue is of particular interest because of its usefulness as an experimental tool for studying the problem of the recovery of crown gall tumor cells. This question will be considered in another section of this article.

Histological aspects [2]

The histogenetic development of a crown gall tumor can best be summarized in terms of the following concept: When a plant is wounded, depending upon the plant and the nature of the wound stimulus, various mature tissues are stimulated into a variety of reactions, including hypertrophic and hyperplastic responses that result in a healing of the wound. Under normal circumstances, upon repair of the wound, the cells which participated in the wound response again return to the quiescent stage characteristic of most mature plant tissue. In the case of crown gall infection this second phase, the return to normalcy, is blocked. Some as yet undefined morphogenetic restraint no longer is applied to, or if applied is no

---

[2] The reader is also referred to an extensive histological and cytological study of crown gall of tomato, pea, and sunflower which appeared after this paper had gone to press (KUPILA 1958).

longer effective upon, the cells of the regenerating tissue. This causes the cells to continue their proliferation in an unregulated manner. An uncontrolled hyperplasia leads to disorganization of the tissues. Disorganization leads to further disorganization. It would be erroneous, however, to envision the tumor-forming process purely in terms of a disorganized cellular proliferation. Many other normal wound healing processes continue to have their counterpart in the developing tumor. Hypertrophy, suberization, and the formation of lignified wound tracheae, for example, may be encountered both in normal wound healing and crown gall formation. It must also be realized that the cells proliferating in the tumor are obviously no longer callus cells, for they no longer behave as normal callus cells but as tumor cells.

The histological sequence of events leading to the production of the tumor varies and is, in part, a reflection of the general characteristics of the tissue potentialities and wound response exhibited by the plant. Thus the types of tissue which, by their proliferation, contribute to the tumor may vary from plant to plant. Most prominent are parenchymatous cells such as those found in the cortex, cambium, and phloem and xylem parenchyma. The pith also proliferates in most plants when the stem has been sufficiently deeply wounded. In some instances specialized tissues such as the epidermis, endodermis, or lignified pericycle fibers may contribute to the hyperplastic mass. A detailed discussion of the histogenesis of crown gall of sunflower will illustrate these points.

The normal response of sunflower (*Helianthus annuus* L.) stems to wounding, *e. g.* by passing a needle through the stem, is typical of most dicotyledonous plants and involves the proliferation of parenchymatous tissues (Stonier 1955). Particularly active are the pith and the cambium. The pith appears to be the most responsive tissue to wounding insofar as wound proliferation may be observed at a greater distance and greater number of cells from the wound edge in the pith than in other tissue. The participation of the cambium in the normal wound response results in the formation of wound wood. Sometimes in deep wounds the proliferating wound cambium may curl inward, extending well into the pith so as to surround the vascular bundles nearest the wound on more than two sides. In such cases, the wound wood may appear as a mass of often radially, and transversely oriented, lignified elements extending into the pith.

When the wound has been infected with crown gall bacteria the development for the first three days is no different from that of the control wound with the exception that it is possible to observe the presence of the bacteria along the edge of the wound cavity, in the intercellular spaces of the pith and cortex, and in the cut xylem. At about this time two other phenomena also begin to manifest themselves, indicating that the tissues are responding to more than just a wound stimulus. The orientation of the cell division appears to be determined not only by the wound surface but also by the presence of the bacteria. In regions removed from the wound, nuclei of cells adjacent to intercellular spaces infected with bacteria may migrate towards the infected space preparatory to forming a sheath of small cells adjacent to the space. Closer to the wound the plane of the phragmosome and subsequent cell wall is no longer solely parallel to the wound surface. A disorientation appears to be taking place. After about four days the cells in the region of a normal wound begin to return to quiescence, in contrast, the infected wound shows signs of increasing rather than decreasing proliferative activity. Thus the expanding cambium continues to proliferate, laying down a large mass of wound wood towards its inside. In cross-section this large mass of well-lignified xylem may in due time be seen to border the wound on both sides. It is this massive xylem "base" which is responsible for much of the woodiness of typical mature sunflower crown gall; it connects with the vascular system of the stem and may be functional.

The expanding pith contributes to the over-all swelling of the stem in forming the crown gall. From time to time tumors may be encountered which appear as large, globose swellings of the stem, lacking the disorganized, discolored appearance of more typical tumors which break through the epidermis. In such cases, more pith than cambial proliferation appears to have taken place. In the pith of such tumors idioblastic wound tracheids may be found.

Whereas it is not difficult to determine the fate of most of the tissue internal to the cambium, that of the phloem is obscure. Very little, if any, phloem tissue remains recognizable in older crown galls. The tissue becomes obliterated, and it is of interest to note that the phloem fibers (i. e. the pericycle or bundle cap) completely dedifferentiate. The walls of the fibers lose their lignin, divide transversely, and enlarge somewhat. As a consequence of this process these cells so resemble the cortical tissue which initially surrounded them that they can no longer be identified.

For a time, the cortex keeps pace with the expanding tumor and in its early stages completely envelops it. The core of the tumor, however, soon breaks through the cortex and epidermis. When this occurs, the outer surface of the tumor becomes covered with a cicatrix consisting primarily of necrotic and suberized cells.

As the tumor grows, the organization rapidly deteriorates. In a relatively short time it usually becomes impossible to trace the cambium from the normal portion of the stem much beyond the lateral edges of the tumor. Instead the tissue fragments into growth centers, the disorganization reflecting, at least in part, the initial degree of infection and resultant growth stimulation exerted by those intercellular spaces which contain bacteria, superimposed upon the stimulus originating from the wound surface. These growth centers expand and in their midst cells mature by hypertrophy or lignification. The presence of these meristematic islands affects the orientation of surrounding cells both in respect to cell division and cell elongation; as the tissue continues to fragment, it becomes more and more disoriented. Ultimately disorganization reaches a stage where necrosis occurs, although in sunflower this is only relatively slight and mostly confined to the surface of the tumor. The probable reason for the absence of extensive necrosis lies in the fact that in sunflower tumors there is considerable formation of lignified vascular elements (mostly tracheids). In typical sunflower crown galls the xylem "base" laid down by the cambium is always recognizable, and even in the more external portions of the tumor the differentiation of tracheids always tends to occur in groups, thereby providing the developing growth centers with vascular tissue. Undoubtedly much of this tissue may be considered to be analogous to the stroma of animal tumors. It is of interest to note that this differentiation of xylem tissue takes place not only in cells which might be considered normal because they were derived from the stimulated cambium but also from cells of regions growing in a rapid and disorganized fashion and which in all probability are derived from tumor cells.

In tissue culture, too, bacteria-free sunflower crown gall tissue which has been cultured for many passages, even years, on media devoid of hormone supplements, still produces tracheids (White and Braun 1942: Struckmeyer, Hildebrandt and Riker 1949).

The development of tumor strands and secondary tumors, referred to in an earlier section, appears to be related to the xylem (Braun 1941). Cell walls in these strands often stain as if acted upon by the bacteria. The response of the xylem parenchyma and pith towards xylem bundles of infected plants is similar to that exhibited by parenchymatous tissues towards an infected intercellular space. The bacteria are known to travel in the xylem of sunflower (de Ropp 1948 c; Stonier 1956 b).

When sterile sunflower tumor tissue which has been cultured *in vitro* for some time is implanted into the stem of a healthy sunflower, the tumor continues to proliferate until the death of the host (White and Braun 1942). The new growth appears only at the site of the graft, while the remaining extent of the wound heals normally. Histological sections, too, indicate that almost all of the new growth is the product of the tumor and that participation by normal stem cells is probably secondary. That is, normal tissue, particularly the cambium, will be stimulated to a limited extent (the cambium laying down the xylem base usually encountered in primary tumors). However, the response is explainable in terms of callusing reactions and the fact that the tumor is known to produce an excess of auxin which would serve to stimulate cambial growth. White and Braun (1942) concluded that there is no evidence that the normal host cells are themselves permanently modified in their behavior by the presence of adjacent tumor cells and that the tumor appears to arise largely, if not entirely, from pre-existing tumor cells.

The picture becomes less well defined, however, when grafts are made *in vitro* between sterile sunflower tumor tissue and freshly isolated normal sunflower stem segments (de Ropp 1947 b, 1948 a, b). The tumor tissue frequently seems to induce a secondary tumor at the graft junction which apparently arises from normal tissue. These induced tumors have a characteristic organization consisting of a spherical woody core surrounded by a cambium-like layer and a layer of thin-walled chlorophyll containing cells outside of which is a loose-celled parenchyma. These overgrowths differ from the cultured tumor tissue in that the latter lacks internal organization, its cells contain no chloroplasts, and nuclear abnormalities are common. Where such "induced tumors" are excised and grown in tissue culture, the organization deteriorates and after about six transfers the tissue once more resembles the typical disorganized tumor tissue used in the initial grafting experiments. De Ropp's data implies, and was so interpreted by him, that some agent passed from the primary tumor to normal cells and transformed those cells to tumor cells. This might well be the case, but other possibilities must be ruled out before this interpretation is accepted. As de Ropp himself stated (1948 b, page 375), on the strength of the evidence it is not possible to conclude definitely that induced tumors arise from normal tissue. He suggested the possibility that the characteristic morphology of the induced tumors might reflect the fact that the tissue consists of a mixture of normal and tumor cells. The fact that the organization disappears at about the sixth transfer might indicate that the tumor cells outgrew the normal cells. Another possibility is that the growth of the tumor and its organization is a reflection of the vascularization, as is known to be the case *in vivo* (White and Braun 1942). Thus it is conceivable that an "induced" tumor arises when the *in vitro* graft results in the tumor tissue fusing with the vascular elements of the host in only one local area. This causes the tumor tissue to proliferate rapidly at such a point, but by virtue of the relative unilateral nutritional supply, physiological gradients may impose some sort of differentiation and organization. A third possibility which could also account for the rare instances in which an induced tumor was observed to develop at the distal end of the grafted stem segment, is that the induced tumors represent habituated tissue. Sunflower stem tissue appears to habituate relatively easily (Kandler 1952; Henderson 1954; Stonier, unpublished results). De Ropp (1947 a, b) has provided several lines of evidence which demonstrate that the sunflower tumor tissue produces an excess of auxin and in an earlier paper (1946) reported that in the tumor-normal tissue grafts the normal tissue does indeed behave as if exposed to a high concentration of auxin. Thus normal tissue might possibly become habituated at the graft juncture and occasionally at the stem surface adjacent to the medium. None of these considerations rule out the possibility, however, that tumorous cells possess an agent which may transform normal cells into tumor cells thereby causing growth by apposition. But neither are there any data which demonstrate unequivocally that this actually occurs.

The preceding descriptions dealing with the histogenesis and other histological aspects of crown gall tumor formation in sunflower serve to illustrate a number of principles which may be generalized to cover other plants as well:

1) The regenerating wound tissue serves as a precursor for the tumor tissue and it is not always possible to distinguish where the former leaves off and the latter begins.

2) The presence of virulent bacteria intensifies the wound response and. whereas wound tissue returns to quiescence after several days, crown gall tumor tissue continues, potentially at least, to proliferate indefinitely.

3) All tissues capable of proliferation may participate in the development of the tumor; this involves, in the case of the lignified phloem fibers, complete dedifferentiation.

4) The extent and pattern of wounding and infection determine the pattern of wound response and on this, in turn, depend many of the consequent histological and morphological properties exhibited by the resulting tumor.

5) Much of the histological disorganization characteristic of crown gall tumors can be accounted for in the following terms: cells surrounding an intercellular space infected with bacteria may respond to it as if it were a wound, providing it is in a region subjected to the general wound stimulus. Such multifocal stimulation causes the orientation of the newly created walls to be in many directions, in contrast to normal wounds in which almost all cell walls are parallel to the edge of the wound. Since the previously existing organization of cells at any given moment determines subsequent organization. disorganization leads to further disorganization as the cells continue to proliferate.

6) Although the tissue becomes disorganized as it forms the tumors, a considerable amount of organization also persists. This is particularly true of sunflower where vascularization is heavy. In the maturing parts of the tumor considerable differentiation takes place and involves not only normal cells but probably tumor cells as well.

7) The growth of the tumor tissue also secondarily stimulates normal tissue into proliferation. There is no clear-cut evidence that tumor tissue may convert normal cells into tumor cells, *i. e.* that tumor growth may occur by apposition.

8) Some, if not all, of the secondary stimulation may be attributed to the excess of growth hormones produced by the tumor tissue. As will be discussed below, certain other histological features may also be ascribed to the presence of an excess of hormones of the auxin type.

In order to consider these principles as being of general applicability to the histogenesis of crown gall tumors, it is necessary to consider them in terms of reports by other workers on other plants.

The importance of wounding to tumor formation will be discussed in another section. That the histological evidence indicates that the regenerating wound tissue serves as a precursor to the tumor tissue has been commented upon by several workers, as has the fact that it is difficult to differentiate the tumor tissue from wound tissue during the early and occasionally later stages of tumor formation. For example, in his review on plant galls, Butler (1930) stated that the general behavior of galls often is an extension of wound responses. Similarly. Berridge, working with tomato and pea stem tumors (1930). suggested that the process of tumor formation is due to healing reactions similar to those which take place in wounded plant tissue. ROBINSON and WALKDEN (1923), describing tumor formation in *Chrysanthemum frutescens*, also concluded that the effect of the bacterial stimulus is to produce a growth very similar in form, structure, and general appearance to callus growth on woody shoots resulting from a response to wounding.

RIKER (1923 a) marked off the water soaked area around infected tomato stem wounds and found that the size of the developing tumor corresponded to this area. He also studied the histogenesis of developing crown galls (1923 b) and subsequently compared and contrasted the tumor development with that of sterile wounds (1927). RIKER pointed to the difficulty of differentiating between wound response and early tumor formation. However, he indicated several important differences. The wound reaction appears to be more localized, while that of the crown gall response is more diffuse and affects a wider area. Whereas the pith responds to wounding by laying down seven to eight cells in a row parallel to the wound edge, the number is greatly increased when the wound is infected. Thus, the swelling caused by the wound never exceeds that of the 4-day tumors since the wound stimulus disappears after five to ten cell divisions, in contrast to crown gall where it persists. Essentially similar observations or implications (when no wound controls were studied) have been made not only in respect to tomato (BUVAT 1945; KLEIN 1952; KUPILA 1956) but in many other plants e. g. *Ricinus* and *Bryophyllum* (COOK 1923), *Pelargonium* (NOËL 1946), and *Vicia faba* L. (KLEIN, RASCH und SWIFT 1953; THERMAN 1956). BUVAT (1945) has suggested that the histological reaction following inoculation consists of two phases: the first involves the "process of cicatrization" following the trauma, the second involves the "process of proliferation" which is caused by the bacteria and produces the tumors. In evaluating such an interpretation it should be recognized that the formation of a cicatrix in itself involves cell proliferation and not until such cell proliferation commences is the tissue transformed to tumor tissue. It is on about the third day following wounding that most of the tumor initiation occurs; it is at this time that one may observe in the tissue surrounding the wound the first wave of cell divisions (BRAUN and MANDLE 1948; BRAUN 1952).

That the similarity between the normal regenerative response and tumor formation may extend beyond the initial 3 or 4 days is indicated by the descriptions of SYLWESTER and COUNTRYMAN (1933). These authors made a comparative histological study of crown gall and wound callus on apple and found that at various stages of development the internal differentiation in graft callus and crown gall is strikingly similar. For example, such features as contorted strands and islands of xylem elements, occasionally organized to resemble a dicotyledonous bundle, are common to both types of overgrowths. The reason for this temporal extension of similar features is obvious. It takes much longer for the cells of a graft callus to cease their proliferative activity than it does for cells stimulated by a wound such as that from a needle puncture. Thus when under natural circumstances the morphogenetic restraints are not applied for a long time to a regenerating tissue, an overgrowth results which resembles crown gall of that species. This would substantiate the concept that it is the second step in the reparative process, *i. e.* the return to quiescence, which is blocked by the tumorous transformation.

Of pertinence to the question of whether the regenerating wound tissue is indeed the precursor of tumor tissue is the question of tissue response and participation. Riker (1927) has stated that all living tissues in tomato which are not lignified join the proliferation but that cells of the cortex exhibit more hypertrophy and hyperplasia than other tissues. Many other authors also find this to be true. Cook (1923) found that the first symptom of crown gall development in *Ricinus* was the formation of a "whorl" of cells in the cortex, usually in close contact with the cambium. He states that the case is similar in *Bryophyllum* except that the connection with the cambium is more pronounced. Similarly, Beck (1949) found that crown galls formed in the cortex of decapitated stems of *Impatiens Balsamina* L. In *Pelargonium* Noël (1946) observed that the cortex and epidermis alone would proliferate and form tumors if the wound was superficial but that in deep punctures the cambium made a considerable contribution. Hamdi (1930) described tumor development in several plants and showed the cambium to participate actively. The cambium or internal phloem is implicated in teratoma formation in tobacco, according to Smith (1917). In tomato Kupila (1956) noted that groups of small cells arise at the inoculation wound near vascular tissue and particularly near internal phloem. She also noted that the epidermal cells first stretch, then divide. According to her observations in pea the development is analogous except that the cortex tends to be less, and the pith more, reactive and the epidermal cells do not divide. Jones (1947) states that in *Rubus* by far the greater bulk of the gall appears to be traceable to the activity of the outer pericyclic cambium. Genevès (1946) concluded that the heterogeneity of crown gall cells is due not to an infiltration of normal tissues by tumor cells but to a cellular hyperplasia and hypertrophy juxtaposed from very early stages. Butler (1930) does not consider pre-existing meristems such as the cambium to be primarily implicated in gall formation, rather he points out that all living cells have the potentiality to react to various stimuli by hypertrophy, hyperplasia, or the development of meristematic tissues. Buvat (1945) has described processes occurring within mature cells which lead to their dedifferentiation into immature, embryonic cells. In *Vicia faba* Klein, Rasch and Swift (1953) observed that tumor formation not only entails proliferation of all parenchymatous tissue but also cell enlargement, dedifferentiation, and division of perimedullary tissue, pericycle fibers, and cortical fiber bundle cells.

In his review on wound healing Bloch (1941) has described the wide range of responses to wounding observed in the plant kingdom including such apparently drastic cell responses as the dedifferentiation of lignified elements. Such observations again reinforce the concept that many of the histological phenomena encountered in crown gall merely seem to reflect an intensified wound response. This suggestion has also been made by Bloch (1952).

If the wound tissue does represent the precursor to the tumor tissue, then some of the properties of the initiating wound should be reflected in the morphology and histology of the mature gall. That this is indeed the case becomes apparent from the fact that there exists a close correlation between the size of the wound and the size of the resulting tumor (see section on the role of wounding). Not only the size but also the histological properties exhibited by a tumor may reflect the properties of the initiating wound. HAMDI (1930) reported that wounds deep enough to penetrate into the pith result in the involvement of all parenchymatous tissues, whereas superficial wounds involve the cortex alone. In tomato, DAME (1938) and HILDEBRAND (1942) obtained tiny galls involving only superficial tissues by rubbing a needle across the stem, thereby injuring only the epidermis and trichomes. Identical results were obtained in *Pelargonium* by NOËL (1946), who reported that in superficial infections produced with a brush the cortex and epidermis may lay down regular rows of cells at the periphery which become suberized. She, too, concluded that the "zone génératrice" delimits the size and participation of tissues in tumor formation.

If there are many similarities between reparative wound callus and tumor tissue, there are also dissimilarities. The most striking of these is the orderly arrangement of cells around a wound in contrast to the disorganized state of the tissues in crown gall tumors. As reviewed by BLOCH (1941, 1952), in the wound the early divisions are generally parallel to the surface of the wound, *i. e.* they occur at an angle of approximately 90 degrees to the direction of gradients from the wound surface. It may be assumed, therefore, that the plane is determined by the shape and influence of the wound itself. Furthermore, in adjacent cells the new walls are very often laid down opposite to each other so that concentric sheaths of cells are formed around the wound. This arrangement is found not only surrounding external surface wounds but also around inner centers of necrosis, *e. g.* infiltrated air spaces or vascular ducts. This picture is valid only when one is dealing with a limited response. As BLOCH indicates and as has already become evident from the descriptions of SYLWESTER and COUNTRYMAN (1933) presented above, in later phases of wound tissue activity the rules of parallelity and opposite positions are often definitely abandoned and cell division may occur in various planes, often coinciding with formation of root or shoot primordia. The plane of cell division, instead of orienting itself in respect to the initial wound surface, now often becomes oriented to the shape and direction of growth of the new cells. BLOCH also indicates that even during early stages of wound healing in the neighborhood of vascular bundles as well as in meristematic pericycle and cambium the cells divide more actively. Thus the presence of vascular bundles also affects the wound response and one may conclude that the previously existing organization of a tissue will bear on the subsequent organization of that tissue.

In the light of these considerations let us now discuss the disorganization observed in the tumor tissue. At least part of this disorganization, as previously indicated, stems from the fact that the tumor tissue continues to grow whereas the wound tissue returns to quiescence. However, this does not account for the fact that the tumor presents a much more disorganized appearance from the beginning, *i. e.* the time the cells begin to respond to the trauma by proliferation. As illustrated in the sunflower stem, the reason for this is twofold: firstly, as discussed above, the infection itself intensifies the traumatic response so that stem tissue reacts more dramatically; secondly, the infected wound area presents a multifocal stimulus to the tissue. That is, the orientation of cells is influenced not only by the surface of the wound but by all sites subjected to the combined action of wound substances and infecting bacteria such as the invaded intercellular spaces and xylem strands. This was first pointed out by RIKER (1923 a, b) who observed that in tomato the bacteria travel in the vascular bundles and may also be found in abundance in the intercellular spaces. Within 2 days cell walls acted on by bacteria acquire new staining properties. Within 8 days cells adjacent to infected intercellular spaces respond by migration of nuclei towards the space, followed by mitosis, thus forming small daughter cells. This response results in the formation of a sheath of small hyperplastic cells around the infected space. Thus much of the regular arrangement found in wound tissue becomes upset. This was also reported by KLEIN (1952) who found that the planes of cell division in tomato crown gall tumors were in all directions, in contrast to the tabular cell arrangement of the control wound. ROBINSON and WALKDEN (1923) and BERRIDGE (1930) found that the cell walls in contact with the bacterial masses appear to be suberized and also suggested that the process of tumor formation is due to healing reactions similar to those which take place in wounded plant tissue. In cane gall BANFIELD (1933) found that strands of the bacteria dissolve the middle lamella of cell walls. He further stated that cells in contact with the bacteria undergo a rapid cytolysis, an observation not encountered in other plants. In *Pelargonium* NOËL (1946) found that a 4-day-old crown gall resembles the wound control except for the reaction exhibited by host cells to the infected intercellular spaces. BECK (1949) observed that the infection of decapitated stems of *Impatiens Balsamina* resulted in the formation of isolated groups of meristematic cells in the cortex near the point of inoculation. The occurrence of meristematic islands in the developing galls has also been noted by JONES (1947) in *Rubus,* KUPILA (1956) in tomato, and SYLWESTER and COUNTRYMAN (1933) in apple.

In view of the intense and multifocal stimulus exerted upon the tissue from the beginning by the infected wound, it is not surprising that the organization of the stem should rapidly deteriorate. What is almost more surprising is that much histological organization remains recognizable. This is undoubtedly a reflection of the fact that the tumor tissue, while *in vivo*, is still subjected to a number of gradients, particularly those involving hormones and nutrition. It is also due, in part, to the extensive differentiation which takes place. For contrary to SMITH's (1921) opinion, there is good evidence to indicate that not only cells derived from what could be secondarily stimulated tissues may differentiate, but also cells initially contained within and presumably derived from tumor tissue. Many authors, working with many different plants, have reported on the occurrence of lignified elements in the highly meristematic portions of the tumors (BECK 1949; HAMDI 1930; JONES 1947; KLEIN 1952; KUPILA 1956; LEVINE 1921, 1924; LOCKE, RIKER and DUGGAR 1938; NOËL 1946; RIKER 1923 b; ROBINSON and WALKDEN 1923; SYLWESTER and COUNTRYMAN 1933; THERMAN 1956). Perhaps even more suggestive is the fact that tumor tissue *in vitro* continues to produce such elements. This was first described in sunflower tissue carried for several passages (WHITE and BRAUN 1942). Subsequently it was found that such sunflower tissue cultured over a period of several years still differentiated tracheids (STRUCKMEYER, HILDEBRANDT and RIKER 1949). Other tissue cultures have behaved similarly, *e. g. Helianthus tuberosus* L. (GAUTHERET 1947 b; KULESCHA 1947). Not only lignification but suberization as well may occur in tumor tissue (NOËL 1946; SYLWESTER and COUNTRYMAN 1933).

Of interest is THERMAN's (1956) recent detailed description of the differentiation of tracheal elements in tumors of *Vicia faba*. The tracheal elements are observable within 9 days post-inoculation. Their differentiation is similar to normal wound healing in *Coleus* as described by SINNOTT and BLOCH (1945). That is, the cytoplasm becomes denser, then forms a definite, usually reticulate pattern along the wall. This pattern is *intercellular* and the secondary wall is laid down along the cytoplasmic pattern. The lignification begins at the thickest spots in the cytoplasmic network, then gradually spreads until the network is completely lignified. At this point the cytoplasm is still present; however, it soon disappears. The nucleus persists somewhat longer but also quickly degenerates. From this description it becomes clear that not only do cells in tumor tissue differentiate, but groups of such cells may do so in unison. This is in accordance with the observations of the workers previously cited in respect to observing lignified elements in the tumor tissue. For not only do lignified elements occur as single idioblasts, but they usually occur as islands and frequently appear to be derived from those meristematic islands which were responsible for much of the initial tumorous growth.

The simultaneous differentiation of cells in a tissue suggests that the cells are exposed to some sort of a gradient. That such a gradient involves the auxinic plant hormone indoleacetic acid (IAA) may be inferred from the facts that tracheid differentiation is dependent on IAA (JACOBS 1952) and that the crown gall tumors are known to be hyperauxinic (see below, and section on tumor physiology). In fact, many of the histological characteristics of the crown gall tumors may be ascribed to the presence of a relative excess of auxin. In 1938 LOCKE, RIKER and DUGGAR first suggested that the cambial activity associated with infected tomato stems reflected the fact that the tumors produced auxin. They based their suggestion on the work of SNOW (1935) who demonstrated that the application of IAA to sunflower seedlings resulted in the proliferation of the cambium. LOCKE, RIKER and DUGGAR (1939) were indeed able to show that an excess of IAA was produced by the crown galls.

JACOBS (1952) has demonstrated that IAA is the limiting factor in the regeneration of xylem strands in wounded *Coleus* stems. Therefore it is highly probable that not only the cambium stimulation but also the formation of the secondary xylem laid down by the stimulated cambium (this structure may be considered to be analogous to the stroma of animal cancers (McKEEN 1954) are attributable to the excess auxin produced by the tumors. It is equally probable that the xylem elements within the disorganized tumor itself are caused by the presence of excess IAA, their disorganized distribution resulting from the fact that within the tumor IAA production is not homogeneous throughout the tissue but is probably irregularly localized. Meristems are known to be a source of IAA production, so that an irregular, localized distribution of IAA production in the tumors is not surprising in view of the irregularly scattered meristematic islands found there. STRUCKMEYER, HILDEBRANDT and RIKER (1949) found that low concentrations of IAA added to the medium caused an increase in the number of tracheal elements of cultured sunflower tumors. In tissue culture, sunflower crown gall tissue tends to differentiate less than in *vivo*. One may account for this phenomenon by considering that in *vivo* the tissue is exposed not only to IAA produced by the tumor tissue but also to that produced by the plant. IAA produced by that portion of the plant above the tumor site would tend to accumulate at the tumor by virtue of the fact that the phloem tissue has become obliterated—the tumor tissue essentially constitutes a partial, perhaps a complete, phloem block. (The idea that the tumor constitutes a phloem block would be in accord with the data of LINK and GODDARD (1951) which shows nitrogenous substances to accumulate in and above the tumors of tomato hypocotyls.) Other tissue culture studies on *Helianthus tuberosus* fragments (KULESCHA 1947) indicate that both crown gall tumors and IAA-induced tissues produce lignified conductive tissue but, whereas the former produces these elements in an irregular fashion, the latter contains vascular masses oriented towards pre-existing tissue. Again, it must be recognized that the histogenesis of a tissue depends not only upon a substance but also upon the distribution of that substance. Thus KULESCHA's observations may be a reflection of the difference between a multifocal stimulation (crown gall tissue) and the stimulation obtained by a more homogeneous unilateral application of the hormone.

The phenomenon of cell enlargement, too, is related to IAA concentrations. At higher (growth inhibitory) concentrations STRUCKMEYER, HILDEBRANDT and RIKER (1949) found that there was increased cell enlargement. In tobacco pith, a system normally devoid of hormones concerned with cell division, JABLONSKI and SKOOG (1954) observed cell enlargement of almost all cells upon the addition of 2×10^{-6} IAA to WHITE's basal medium. Cell enlargement in this system is accompanied by cell division when a medium containing an auxin is supplemented with coconut milk. Such a cell division factor was also found in crown gall tumor tissue (BRAUN and NAF 1954; STEWARD, CAPLIN and SHANTZ 1955). Subsequent investigations indicated that these naturally occurring cell division factors could be replaced with 6-furfurylaminopurine (MILLER et al. 1955 b). From these observations it becomes apparent that if a cell were placed in a position such that it would be exposed to a low concentration of the factors limiting for cell division but a high level of IAA, it would tend to enlarge rather than divide.

Finally, it should be pointed out that not only may IAA produce overgrowths on certain plants (BROWN and GARDNER 1936) but the histology of these self-limiting growths bears a striking resemblance to crown gall tumors (KRAUS, BROWN and HAMNER 1936; HAMNER and KRAUS 1937; BECK 1949; APPLER 1951).

As has already been discussed above, the hyperauxinic state of the tumor tissue may cause normal tissue in the vicinity of the tumor tissue to become secondarily stimulated. Furthermore, this condition presents to the observer a histological gradient starting with 1) quiescent normal tissue, to 2) dividing normal tissue which is still organized, to 3) dividing normal tissue where the organization may break down, to 4) disorganized tissue which is undoubtedly tumor tissue. Except where one is dealing with grafted tumor tissue (e. g. WHITE and BRAUN 1942), the demarcation line may be indistinct and it is not surprising that some authors, notably SMITH (1922) concluded that growth occurs by apposition. Perhaps the most conclusive evidence negating this contention consists of the brilliant experiments reported by JENSEN (1910, 1918) who grafted together various combinations of tumors and normal roots of sugar beet (white tissue), garden beet (red tissue), and yellow mangel (yellow tissue). In each case the color of the tissue served as a marker and it became apparent that growth of the grafted tumor is in an outward direction and that the growth is from the tumor, not the normal tissue. Like many another brilliant experiment, this one was at least a quarter of a century ahead of its time and the results were completely ignored.

JENSEN's experiments, however, may not be entirely applicable to some other plants. As discussed in the detailed histological analysis of sunflower crown gall, in view of DE ROPP's data the possibility must be recognized that under special conditions normal tissue might become habituated and thus move in a neoplastic direction. Similar reservations must also be kept in mind for the closely related Jerusalem artichoke (H. tuberosus) where similar conditions prevail (CAMUS and GAUTHERET 1948).

Cytological aspects

The difficulty of differentiating between a process which involves growth by apposition and one which involves only secondary stimulation stems from the fact that it is not always possible to recognize a tumor cell as such by looking at it. Investigations dealing with the morphological aspects of tumor cells, whether they are concerned with the nucleus or the cytoplasm, have so far been singularly unhelpful in that respect. That abnormal nuclear behavior often is associated with crown gall tumor cells has been observed by many workers since the initial report by SMITH, BROWN, and McCULLOCH (1912), including SMITH himself (1916 c, and subsequent papers). Auler (1924), Braun (1947 b), Kostoff and Kendall (1932), Kupila (1956), Levine (1925, 1929, 1930, 1931), Milovidov (1930), Riker (1927), Therman (1956), and Winge (1927). These reported abnormalities may be placed in two categories: *1*) those involving obvious morphological abnormalities such as amitotic divisions, formation of giant, lobed or otherwise abnormally shaped nuclei, or multinucleate cells, and 2) those involving polyploidy. The relatively late appearance in tumor development of the first category of abnormalities implies that they reflect an effect rather than a cause of abnormal proliferation. No author has considered these abnormalities to be the direct cause of the neoplastic state although it has been considered symptomatic of this state. In the second case, *i. e.* that of polyploidy, it is not possible to dismiss, *a priori*, this condition as a secondary phenomenon. In fact, some authors have attempted to fix the primary cause of the neoplastic state in terms of such aberrant nuclear behavior. Winge (1927) postulated that the tetraploidy which he observed in tumors of sugar beets might satisfactorily explain the "growth energy" exhibited by the tumor tissue and that there existed an upset in the number of genes favoring growth over the number concerned with the regulation of growth. Some authors, however, have been unable to observe any polyploidy over and above that encountered in normal tissue (Buvat 1943; Garrigues 1943; Klein, Rasch and Swift 1955). It must be remembered that polyploidy may be frequently encountered in the nuclei of mature plant tissues (see reviews by Lorz 1947; D'Amato 1952). Polyploidy in normal plant tissues does not, itself, appear to appreciably affect the formation of tumors since Grasselli and Cova (1956) were unable to observe any significant difference between tumors produced on diploid tomato plants and those produced on a tetraploid race of the same species.

Other investigations indicate that polyploidy is not a cause of the neoplastic state but more likely the product of it. Levine (1931) found that in young tumors of tobacco and beet the outer, actively growing zone contains normal diploid mitoses, while the inner mature zone exhibits polyploidy. Kostoff and Kendall (1932) inoculated a large number of tomato plants internally. When the tumors reached pea size, the plants were decapitated three to five centimeters above the tumors. As a result, secondary shoots were produced, a small number forming from the tumorous regions. These shoots were removed and planted in soil; five of seven shoots rooted. It could be shown that one of the five shoots was tetraploid. According to Winge (1927) a tetraploid tissue derived from a tumorous region ought not to grow into a normal plant. Kraus, Brown and Hamner (1936) showed that the formation of indoleacetic acid-induced overgrowths on bean plants involves a very great speeding up of nuclear divisions, such divisions frequently resulting in multinucleate cells. More recently Naylor, Sander and Skoog (1954) reported that in isolated tobacco pith fragments, a system requiring for growth by cell division not only auxinic hormones but also the coconut milk factors, the application of physiological concentrations of IAA alone results in mitosis and cell enlargement without any cell division. Mitoses are observable within 42 hours, with a maximum number of mitotic figures at 70 hours, after the start of the cultures. Cell enlargement, on the other hand, is first observed after 72 to 96 hours. Moreover, many of these IAA-induced mitoses result in polyploid nuclei, and after 100 hours in culture many cells are multinucleate or contain lobed or otherwise abnormal nuclei. These artificially stimulated multinucleate pith cells do not continue to grow after 2 or 3 weeks. It is clear from these observations therefore that there is no direct correlation between polyploidy and the neoplastic state since nuclei of both tumor and normal tissue may, or may not, be polyploid. It is also clear that all nuclear abnormalities observed over and above those which take place in non-neoplastic tissue can be accounted for in terms of the abnormal growth hormone physiology characteristic of the tumor tissue.

In shifting attention from the nucleus to the cytoplasm, the structure of the cell again fails to define what constitutes the neoplastic state. Aside from the observation that the cytoplasm becomes denser and stains more heavily, a phenomenon well known to occur in cells preparing to divide, there also occur certain changes in the microscopic particulate structures of the cell. The first report of such a change in tumor cells came from SMITH (1912 a) who reported that with the use of a gold chloride stain he could observe "bacteria" in the cells. These bodies, according to SMITH, occupy only a small portion of the cells and occur in the form of rods two to three times as long as broad, in short chains, short filaments, and Y-shaped or variously branched forms. Subsequently he admitted, however, that these bodies "which at one time I identified as bacteria" were probably mitochondria (SMITH 1920, page 466 and Figs. 331, 332). Since that time some authors have reported that they were unable to observe any conspicuous changes in the mitochondria (BANFIELD 1935; MILOVIDOV 1930), while others did. Perhaps the most extensive studies are those of BUVAT (1943, 1945) on tomato stem cortex. According to BUVAT, at the time of the first mitosis the new activity of the cell manifests itself by a change in the "chondriosomes." These bodies swell and form a non-staining central region, the precise nature of which is not known (it is not starch). Towards the end of the first (wound) phase, the colorless region is resorbed again. Subsequently the swollen chondriosomes undergo repeated division until finally the cell contains granular mitochondria and short rods resembling those of embryonic cells. BUVAT also describes the dedifferentiation of chloroplasts.

A third category of observations involves the deposition of certain substances in the cytoplasm, notably oxalates and tannins (KUPILA 1956; MAGROU 1927; MANIGAULT 1953; MILOVIDOV 1930; PINOY 1925; RIKER 1923 b, 1927; RIKER and BERGE 1935; SYLWESTER and COUNTRYMAN 1933; THERMAN 1956). It is interesting to note that one of the few differences between callus overgrowths and crown gall tumors observed by SYLWESTER and COUNTRYMAN involves the formation of tannins. No differences were noted by these authors in respect to cellulose, pectin, lignin, and gums, but crown gall tissue contained tannin whereas the callus tissue did not. These deposition products generally do not appear until the tumor tissue has been proliferating for some time and may therefore be considered as a secondary phenomenon.

As to the question of whether or not the reported changes of the cell particulates represent primary or secondary changes, *i. e.* whether these changes confer upon the cell the neoplastic state, the following points need be made. *1*) As has been indicated in the section dealing with histology, during the earliest phases one cannot, simply by looking at the cell itself, differentiate between a cell responding to a wound stimulus, which will after a few divisions return to quiescence, and one which has become transformed to a tumor cell and which will continue to proliferate. *2*) In both cases parenchymatous cells undergo a certain amount of dedifferentiation, *i. e.* they once more resemble more closely the dividing cells of the meristem. *3*) This process of dedifferentiation may affect not only the cytoplasm and cell walls but also the cell organelles. The nucleus rounds off, as described by Therman (1951); the visible particulates exhibit a behavior as described by Buvat (1945) and more recently by Fogelberg, Struckmeyer and Roberts (1957). The reason for putting the changes reported for the mitochondria in the category of cell dedifferentiation stems from the report of the last authors mentioned, who observed such changes not only in mature pith cells responding to crown gall infection but also in cells responding to the application of growth-promoting chemicals of the auxin type. Thus it becomes apparent that such observed changes may also occur when a non-tumorous cell prepares to divide.

These considerations, however, should not be construed to imply that an alteration of the cell particulates might not take place. It merely emphasizes that the observed changes appear not to be a reflection of the neoplastic cell transformation. One should keep in mind that if the neoplastic transformation involved the alteration of a single enzyme or enzyme system, the chances of observing this by looking at a mitochondrion are about as great as picking up a gene mutation by looking at a chromosome.

Physiological aspects

A. The two-phase concept of tumor development

In attempting to analyze a complex series of events such as those that occur during tumor inception and development, it is often helpful to subdivide, in so far as that is possible, the total event into a series of contributing events each of which is essential for the consummation of the completed process. The first real attempt to deal with part events in crown gall tumor formation resulted from the studies of Braun and Laskaris (1942). In these experiments tomato plants were decapitated and all axillary buds were removed in order to eliminate, as far as possible, the major sources of a naturally occurring growth hormone, auxin. It had been found by earlier workers (Hendrickson, Baldwin and Riker 1934) that isolates within strains of the crown gall bacterium may show different degrees of virulence. Some isolates are capable of regularly initiating large rapidly growing tumors (Fig. 2, A) while certain sister cell isolates produce small very slowly growing tumors at points of inoculation (Fig. 2, B). LOCKE, RIKER and DUGGAR (1938) demonstrated that these normally slowly growing tumors could be stimulated to rapid growth when tumors initiated by the virulent culture developed above those initiated by the attenuated strain. It was found, further, (BRAUN and LASKARIS 1942; THOMAS and RIKER 1948) that when the small overgrowths

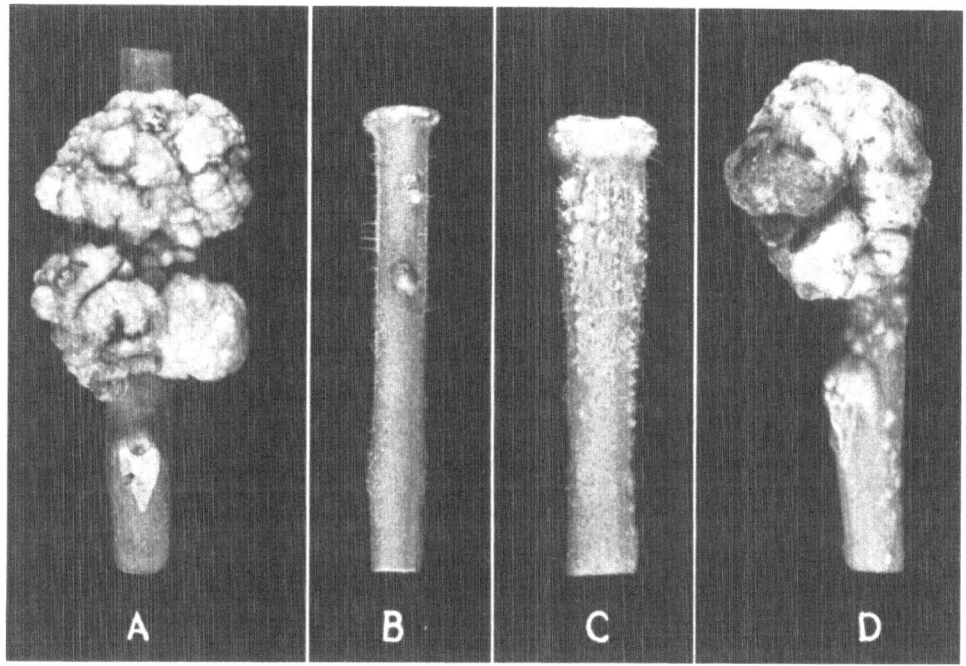

Fig. 2. Tomato stem inoculated with A, a virulent strain of the crown gall bacterium; B, an attenuated strain; C, punctured with a sterile needle and treated at the cut stem surface with a synthetic hormone of the auxin type; D, an attenuated strain and treated with a synthetic hormone of the auxin type.
(Photographs by J. A. CARLILE.)

were supplemented at a distance with growth substances of the auxin type such as naphthalene acetic acid or indole butyric acid, the normally slowly growing tumors also expanded rapidly and were comparable both in size and rate of development to tumors initiated by the virulent strain of bacteria (Fig. 2, D). Bacteria isolated from such artificially stimulated tumors were still attenuated, indicating that the treatment did not bring about an increase in the virulence of the attenuated organisms. The growth-promoting substance caused some cell proliferation to occur at the point of application in uninoculated control plants but this growth was of a self-limiting type (Fig. 2, C). When fragments of the artificially stimulated tumor tissue were implanted into a healthy host containing a functional apical bud, they gave rise again to typical crown gall tumors. Fragments of normal tissue stimulated to proliferation by a growth substance, when grafted into a healthy host fused with the host but did not develop into tumorous overgrowths.

As a result of this study it was concluded that essentially three factors are involved in accomplishing full tumor development. Needed are 1) the tumor-inducing principle associated with the bacteria which serves to bring about conversion of normal cells to tumor cells; 2) susceptible host cells which are capable of responding to the carcinogenic influence of the bacteria; and 3) a hormonal effect. Cells transformed by virulent bacteria apparently generate optimal or near optimal amounts of a growth-promoting substance(s) for their continued rapid proliferation. Those transformed by an attenuated culture have had their requirements for rapid growth only partially satisfied for this substance(s).

As a result of this study it was also hypothesized that tumor formation takes place in essentially two distinct phases. In the first phase normal cells are converted to tumor cells which as yet do not develop into a neoplastic growth. The second phase, according to this concept, involves those factors that are concerned with the continued growth of the altered cells. The virulent bacteria accomplish both phases, whereas the attenuated culture brings about essentially only the first phase. This study emphasized, then, a need for recognizing a distinction between those factors that render the cells neoplastic and those that affect their subsequent behavior. As will be discussed below, Klein and Link (1952) have further subdivided these phases.

B. The role of the bacteria in tumor formation

That the crown gall bacteria are necessary only to initiate tumors was first implied by Jensen's studies (1910, 1918). This worker grafted presumably sterile fragments of sugar beet tumor tissue to healthy beet roots and obtained tumorous overgrowths from the implanted tissue. Due, however, largely to the influence of Erwin F. Smith, the concept that crown gall was a bacterial hyperplasia was not discarded until the early 1940's when it was demonstrated unequivocally (Braun 1941; White and Braun 1942; Braun and White 1943) that sterile tumor tissue could be obtained from secondary tumors of the sunflower. Such tissue possessed the capacity for unlimited growth both in situ and in vitro. Subsequently bacteria-free crown gall tumor tissue was isolated from primary tumors of many different plant species (White 1945; de Ropp 1947 c; Morel 1948; Hildebrandt and Riker 1947, 1949). Bacteria-free crown gall tumor tissue is commonly characterized by a capacity to grow profusely and indefinitely on a simple chemically defined culture medium, such as White's (1954) basic medium, that does not support the continued growth of normal cells of the type from which the tumor cells were derived. When fragments of such sterile crown gall tumor tissue, but not thoroughly ground tumor cells, are implanted into a healthy host, they develop again into tumors that are indistinguishable from those initiated originally by the bacteria. These studies indicate then that the bacteria induce a profound and heritable change in the subsequent behavior of the affected cells. They give no indication, however, whether the bacteria accomplish the transformation in short periods of time or whether long periods of irritation are required to bring about the cellular change. In order to study this question, an experimental method was used (BRAUN 1943, 1947 a, 1951 b) that permitted the selective thermal killing of the bacteria at any desired time following their introduction into a host. This method did not affect the behavior of the host cells transformed to tumor cells prior to the time that the inciting organisms were destroyed. With the use of this method it was shown that normal plant cells may be converted to tumor cells as early as 34 hours after the bacteria are introduced into a host. Tumors initiated at this early period grew very slowly and remained quite small even after prolonged incubation. When, on the other hand, the bacteria were permitted to act on the host cells for 4 days before being destroyed selectively, the resulting tumors were comparable both in size and rate of development to those found in the inoculated but unheated control plants. Between these two extremes the size and rate of growth of the resulting tumors increased progressively, depending on the length of time the bacteria had been allowed to act. These studies indicate, then, that a factor of considerable biological interest passes from the bacteria to the host cells and brings about a complete and heritable change in the subsequent behavior of the affected cells. This as yet uncharacterized factor has been called the tumor-inducing principle. These experiments have been fully confirmed and extended (THEIS, RIKER and ALLEN 1950) to show that, in addition to high temperature, a relative humidity of more than 60 per cent is essential to accomplish the thermal destruction of the bacteria.

Cells altered to tumor cells in a 34-hour period have been found to grow very slowly in culture as well as in the host as compared with the growth rate of cells altered by the bacteria in a 4-day period. Since these characteristic growth patterns have been maintained over a period of more than 10 years, it was suggested (BRAUN 1951 b) that the alteration of normal cells takes place gradually, leading in a 4-day period to a completely autonomous rapidly growing type of cell. Cells altered in a 34-hour period represent, then, a lower grade of neoplastic change than do cells altered by the bacteria in a 96-hour period.

This method of subjecting plants to sufficiently high temperatures to destroy the bacteria is a rather drastic procedure and a milder method was subsequently developed to study more adequately the conditions that are critical to the transformation of normal cells to tumor cells. This method was based on an observation originally made by RIKER in 1926. RIKER (1926) found that tumors produced on tomato plants held at temperatures of 28⁰–30⁰ C. were distinctly inferior in size to tumors initiated at lower temperatures. No tumors at all were produced at temperatures above 30⁰ C. This was true despite the fact that both the host and the bacteria grew well at temperatures up to 32⁰ C. The underlying reason for this interesting finding remained obscure for more than 20 years. By applying the two-phase concept of tumor formation to this and similar systems, it was shown (Braun 1947 a) that a temperature of 30⁰ C. or more prevents the transformation of normal cells to tumor cells but does not interfere with multiplication of the tumor cells once the cellular alteration has been accomplished. Thus the inception process can be brought to an abrupt and complete halt at any desired time following inoculation of the plants with bacteria by simply placing and holding treated plants at or above 30⁰ C. The size and rate of growth of tumors that subsequently developed at 30⁰ C. reflected the degree of cellular alteration that had occurred at a lower temperature of 25⁰ C. up to the time the plants were placed and held at or above 30⁰ C. With the use of this method a number of facts relating to the process of tumor inception in crown gall were established. The inception period was very accurately defined for several plant species (Braun 1947 a; Braun and Mandle 1948; Segrétain 1949). In accord with the earlier results in which the bacteria were killed, it was found with this method that there is a lag of somewhat more than one day between inoculation of the plant with the bacteria and the first evidence that the cellular alteration has occurred. Tumors of the largest most rapidly growing type were initiated only after the bacteria had acted on the host cells for 72 hours or more, while the host cells were for the most part no longer susceptible to alteration 5 days or more after wounding. The latter was true despite the fact that many virulent bacteria were in intimate contact with the host cells after the fifth day. In the hosts studied, cellular transformation appeared to be dependent upon a more or less unconsolidated wound. Histological studies (Braun and Mandle 1948) concomitant with the time-temperature experiments indicated that it is just before or during the earliest stage of active healing that normal cells are converted to tumor cells of the most rapidly growing type. The inability of the bacteria to accomplish the cellular alteration at or above 30⁰ C. was found not to be the result of physiological disturbances in the plant as a whole (Braun 1947 a) but is dependent upon environmental conditions that exist in the immediate vicinity of the site of inoculation. It was found, moreover, that the gross wound healing response, bacterial multiplication, and the growth rate of tumor cells after the cellular alteration had been consummated were slightly enhanced at 32⁰ C. when compared with responses obtained at 26⁰ C.

The minimum period necessary for the bacteria to alter normal plant cells to tumor cells could be reduced from 34 hours to 10 hours if the bacteria were allowed to establish themselves in a host for 24 hours at 32⁰ C. prior to being placed for 10 hours at 26⁰ C. Periods of less than 10 hours were inadequate, whereas progressively larger tumors resulted when the period of time at 26⁰ C. was increased from 10 to 30 hours. Since tumors initiated in 10 hours at 26⁰ C. following a 24-hour incubation period at 32⁰ C. were essentially the same size as were tumors that ·developed when a total incubation period of 34 hours at 26⁰ C. was given, it was concluded that the first 24 hours after inoculation either represented an incubation period necessary for the bacteria to establish themselves in a host or that it represented a period necessary to render the host cells susceptible to transformation.

In further experiments (Braun and Mandle 1948) plants were alternately subjected to temperatures of 25⁰ C. and 32⁰ C. following an initial incubation period of 24 hours at 32⁰ C. Five 6-hour periods at 25⁰ C. were alternated respectively with 1, 2, 4, or 6-hour periods at 32⁰ C. It was found in these studies that when five 6-hour periods at 26⁰ C. were alternated with 1-hour periods at 32⁰ C., large tumors formed. As the intervals that the plants were held at 32⁰ C. were increased in relation to the time they were held at 26⁰ C., the size and rate of development of the resulting tumors decreased. When total exposures at 32⁰ C. approached or equaled those at 26⁰ C., occasional small tumors or, in most instances, no tumors at all developed. On the basis of these results it was suggested that: a) the tumor-inducing principle is cumulative as evidenced by the fact that, whereas a single 6-hour exposure at 26⁰ C. was inadequate to accomplish the alteration, a series of such exposures interrupted by short periods at 32⁰ C. permitted the alteration to occur; b) the tumor-inducing principle is inactivated at the higher temperature. If no inactivation had occurred, the size of tumors resulting from the different exposures at 32⁰ C. should have been much more equal because in each instance the same total of 30 hours at 26⁰ C. was given; c) the tumor-inducing principle was inactivated or destroyed at 32⁰ C. at about the same rate that it was produced at 26⁰ C.

C. The role of the wound in tumor inception

Very early in the study of the crown gall disease it was recognized that a wound is necessary if a primary tumor is to form. The essentiality of a wound was initially conceived of in terms of an entrance site for the invading bacteria (Smith, Brown and Townsend 1911). Since Smith believed the bacteria to be intracellular, the wound was thought necessary to expose the protoplasts. Magnus (1915) vaguely implied that wound callusing was involved and that tumor responsiveness of various plant species appeared to be a reflection of their respective callusing ability. That exposing living protoplasts to the bacteria was not essential was demonstrated by Riker (1923 a) who inserted a red hot needle, one-half hour post-inoculation, into the puncture wound of a tomato, thereby killing all of the cells whose walls had been ruptured as a result of the initial inoculation. The size of the resulting tumor was not affected by this treatment. Instead, Riker reported, the size of the tumor coincided with the sap-soaked area around the wound. It has generally been found, in accordance with the above observation, that the ultimate size of the tumor is within limits proportional to the size of the initial wound (Hamdi 1930, Hildebrand 1942; Levine 1923 a, 1936). The addition of plant sap to small wounds appeared to favor the development of somewhat larger tumors than would otherwise have developed (Hamdi 1930; Hildebrand 1942). Klein (1955) has recently shown that carrot phloem discs that respond poorly to inoculation with crown gall bacteria can be made to react strongly by applying expressed juice from normally strongly reacting carrots. All of these studies suggest the possible role of wound sap in the genesis of tumors. Two possibilities, therefore, present themselves as to the role of wounding in tumor initiation: *1)* the wound is necessary to condition the host cells and thereby make them susceptible to transformation, and *2)* the wound sap is essential not only as a substrate for the bacteria but more specifically to permit the bacteria to elaborate the tumor-inducing principle from some precursor present in the plant sap. These possibilities need not be mutually exclusive.

The specific role of a wound in the inception of the crown gall tumor was recently found (Braun 1952). It was demonstrated in these studies that host cells must be conditioned by the stimulus of a wound before they become susceptible to alteration by the tumor-inducing principle. This effect was demonstrated by comparing the response of wounded tissue that had been permitted to heal for 48 hours prior to inoculation with the bacteria with the response obtained when the bacteria were introduced directly into previously unwounded tissue. In both instances the bacteria were allowed to act for only 24 hours at 25° C., a period which is not in itself sufficient to permit the cellular transformation to occur when the bacteria are introduced into previously unwounded tissue. In these experiments plants that were wounded 2 days prior to inoculation with bacteria developed large tumors, while those inoculated at the time of wounding showed no tumorous reaction. Conditioning of the host cells was shown, moreover, to take place gradually, reaching maximum susceptibility between the second and third days after a wound is made and declining again as wound healing progresses toward completion. If, therefore, the host cells are not adequately conditioned, as appears to be the case in the early and late stages of wound healing as well as in most normal cells not under the influence of a wound, then the cellular transformation will not occur despite the fact that many virulent bacteria are in intimate contact with such cells. This process was found to take place independently of the bacteria at temperatures of both 25° and 32° C., although the cellular alteration was accomplished only at the lower temperature.

As a result of a wound, profound changes in the physiological behavior of essentially resting cells in the region of a wound occur (Bloch 1941). These modifications include changes in the permeability of the cell membranes, an increase in cycloses as well as respiratory modifications. Doubtless other as yet uncharacterized changes also occur. The specific role of such changes in the conditioning of cells for transformation remains at present unknown.

Segrétain (1949) has postulated that heat-labile substances produced by the cells during cicatrization may play a role in tumor inception. No evidence favoring this view has been' presented and it should be recalled that the wound healing response as such is not affected by a temperature of 30° C. Furthermore, the addition of plant juice from conditioned stem sections to the bacteria at the time of inoculation into non-conditioned plants fails to produce tumors if held at 25° C. for only 24 hours (Braun 1952).

The possibility remains, however, that a substance is released which is involved not in wound healing *per se* but in tumor induction. Indeed, KLEIN (1954) has postulated that wound juices not only condition the host cells but also provide the bacteria with a substance(s) which is essential to the synthesis of the tumor-inducing principle by the bacteria. Unfortunately, it is not possible to resolve this question at this time because one cannot as yet obtain conditioned cells without plant sap. Alternatively, should it become possible to isolate and positively identify the tumor-inducing principle, then factors involved in its production could be ascertained with exactitude.

D. The possible role of bacterial auxin as a "cocarcinogen" in tumor genesis

The etiological role of the tumor-inducing principle responsible for the transformation of normal cells to tumor cells and of conditioned host cells has now been clearly established in tumor genesis. Recently KLEIN and LINK (1952) have reported results which they have interpreted to mean that a third factor, bacterial auxin, is implicated in the transformation process. According to these workers the affected cells acquire as a result of the action of the tumor-inducing principle only the potentiality for autonomous growth but not the capacity for rapid duplication. Bacterial auxin is needed as a "cocarcinogen" to complete the transformation process and promote the rapid multiplication of the tumor cells. The evidence for this concept was based upon experiments similar to those described earlier in this discussion (BRAUN and LASKARIS 1942). An attenuated culture of crown gall bacteria was inoculated into tomato plants and lanolin paste containing indole butyric acid was applied to the cut stem ends at a distance from the points of inoculation. The pastes were allowed to remain for periods ranging up to 240 hours. If the auxin-containing paste was removed from the cut stem surface 36—48 hours after inoculation and application of the paste, it had no effect on tumor development. Between 48 and 60 hours after inoculation the eventual size of the tumors was a function of the length of auxin presentation. Subsequent to 60 hours there was found to be no further need for supplementary growth substance for there were no significant size differences among tumors formed under conditions of extended auxin-presentation time. These workers reported, further, that the incipient but unpromoted tumor cells remain positively predisposed to auxin for some time. When the auxin paste was applied as late as 10 days after inoculation of tomato plants with attenuated bacteria, large tumors were produced. However, when 15 or more days elapsed between inoculation and promotion, the resulting tumors contained root primordia. This was interpreted to mean that the host plant had to some extent gained control over the development of the neoplastic tissue.

In order to establish the concept that bacterial auxin acts as a cocarcinogen and thus completes the cellular transformation, it would appear necessary to show that the growth-substance stimulated cells altered by an attenuated culture reached and retained indefinitely the high degree of neoplastic change found in cells altered by the virulent culture. It is possible to obtain a pronounced stimulation of cells transformed by the attenuated culture in tomato plants by inoculating a virulent culture at a distance of five centimeters above the attenuated culture, as was first shown by Locke, Riker and Duggar (1938). As the tumor initiated by the virulent culture expands, biologically active substances diffuse in a basipetal direction in the stem and stimulate cells altered by the attenuated culture to rapid growth in much the same manner as do the synthetic auxins. The results obtained with this technique, however, are more nearly reproducible than are those obtained when synthetic growth substances are used. With the use of this method it was possible to demonstrate (Braun, unpublished results) that tumors initiated by an attenuated culture continue to develop actively only as long as a source of growth substance diffusing from the tumor initiated by the virulent culture is present. Once the source is removed the growth rate of cells altered by an attenuated culture again slows down to a low level, indicating that the cells themselves have not acquired the capacity for rapid growth as a result of their exposure to the growth substances. It was found, in contrast to the results reported by Klein and Link (1952), that tumors initiated by an attenuated culture can be stimulated to rapid growth as long as 6 weeks after inoculation. Results such as these suggest that crown gall tumor cells themselves acquire, as a direct result of the action of the tumor-inducing principle, a capacity for producing in greater than regulatory amounts a growth substance of the auxin type. According to this interpretation, then, cells altered by the attenuated culture represent a lower grade of neoplastic change than do cells altered by a virulent culture. By supplying such normally slowly growing tumor cells with an external source of growth substance of the type that is normally limiting for the rapid growth of such cells, those cells are being supplemented with a biologically active material that is elaborated in abundance by cells altered by the virulent culture. Recently Link (Klein and Link 1955, page 211) has concurred with the interpretation of Braun and Laskaris (1942) and Braun (1954 a) that induction completes the alteration of a normal cell to a tumor cell.

E. The growth-substance metabolism of crown gall tumor cells

The uncontrolled or autonomous growth of tumor cells generally appears to reflect a deranged metabolism in those cells. An understanding of the problem of autonomy will in all likelihood, therefore, depend upon a characterization of the biochemical lesion(s) that is responsible for this altered metabolic state. Since tumor cells appear somehow to have acquired the capacity for autonomous growth as a result of uncontrolled growth and

---

Fig. 3. Histological sections of tobacco pith parenchyma tissue: *A*, untreated. *B*, treated with naphthalene acetic acid. Note that the pith cells have enlarged greatly without dividing. *C*, treated with 6-furfurylaminopurine, a factor limiting for cell division in this system. Note that this compound when used by itself does not encourage either cell enlargement or cell division. *D*, treated with naphthalene acetic acid and 6-furfurylaminopurine. A rapid growth accompanied by cell division results from the synergistic effect of these two growth substances.
(Photographs by J. A. Carlile.)

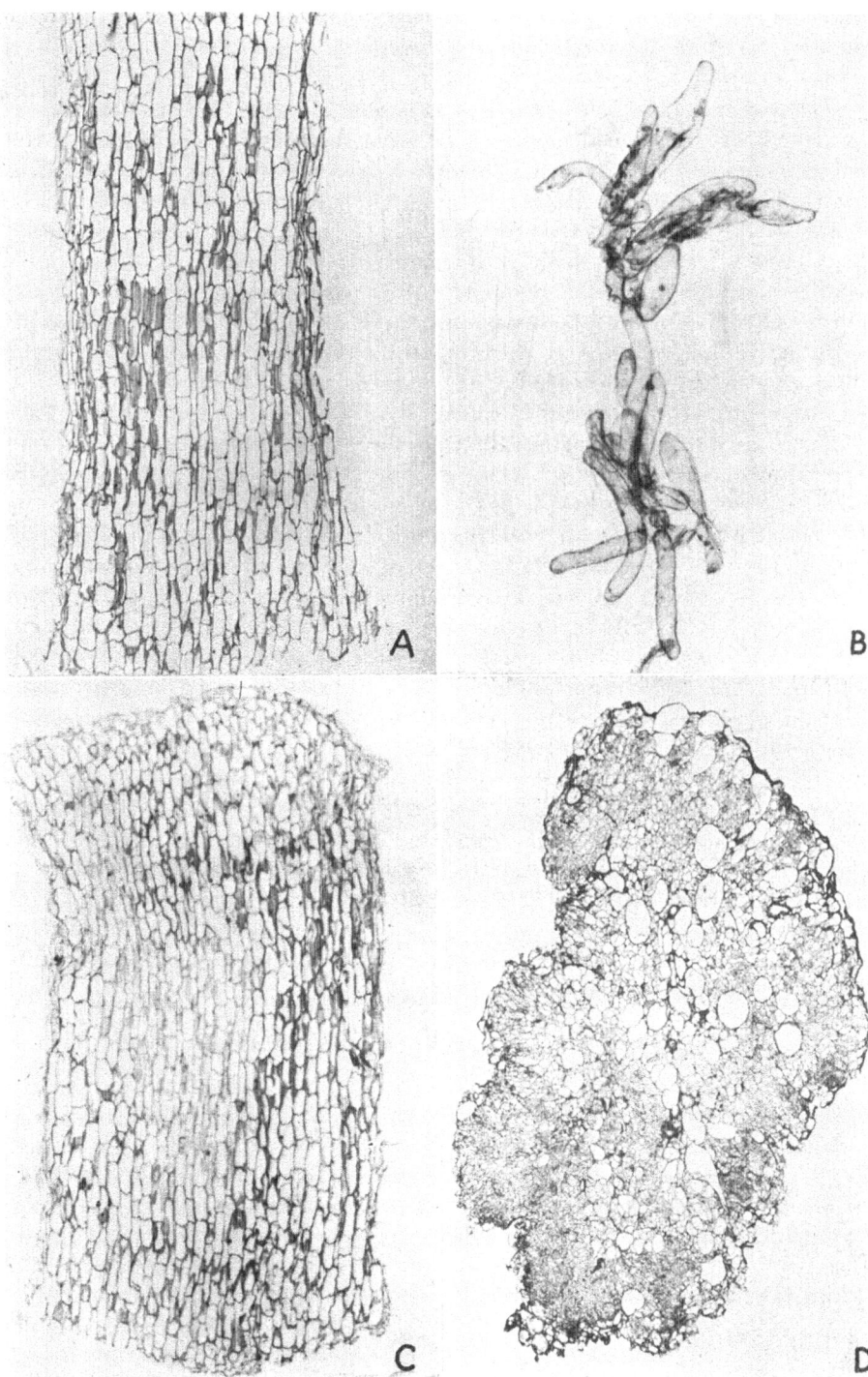

division, attempts have recently been made to explore this problem through an analysis of the factors concerned specifically with growth accompanied by cell division.

Growth in all higher organisms commonly results either from an enlargement of their constituent cells or from the combined processes of cell enlargement and cell division. These two processes appear to be dependent for their development in higher plants upon specific substances that can be synthesized by the plant cells themselves. The auxins, of which indoleacetic acid is the best known naturally occurring representative, induce pronounced cell enlargement in tobacco pith parenchyma tissue to the exclusion of cell division (Jablonski and Skoog 1954). In this system biologically active factors present in such naturally occurring substances as the fluid endosperm of the coconut are required in addition to an auxin to encourage growth accompanied by cell division (Jablonski and Skoog 1954). Without an auxin the factors limiting for cell division are ineffective in promoting this process in tobacco pith parenchyma cells (Fig. 3). Recently 6-furfurylaminopurine, which has been assigned the trivial name kinetin, has been found capable of replacing the coconut milk factors in promoting cell division in this system (Miller et al. 1955 a, b). It is thus possible with the use of certain specialized plant cell types such as tobacco pith to delimit under fully controlled conditions the factors concerned with cell enlargement and those concerned with growth accompanied by cell division. Plant cells require, therefore, in addition to water, inorganic and organic nutrients, and various cofactors, other specific substances which determine whether a plant cell can fully exploit its inherent potentialities to enlarge and divide. The growth-substance synthesizing systems concerned with the production of the cell enlargement and cell division factors are, however, precisely regulated in all normal plant cells. Since these growth substances appear to be concerned specifically with growth and cell division, attempts have recently been made to learn the fate of these two as well as several other growth-substance synthesizing systems during and following the conversion of normal plant cells to tumor cells in the crown gall disease (Braun 1956 a, b; 1957 a, b).

1. Auxin metabolism of the crown gall tumor cell

Of the hormones concerned in growth of higher plants, the auxins have been studied most intensively over the years. The maintenance of a balance between the production and destruction of auxin can be regarded as a characteristic of the normal metabolism of a higher plant. Much evidence of both a direct and indirect type has accumulated over the past two decades to indicate that this balance is not attained in a plant tumor such as crown gall. That these tumors are hyperauxinic and elaborate auxin in greater than regulatory amounts is evidenced by the following findings.

Host responses characteristic of high auxin levels such as pronounced epinasty of the leaf petioles, formation of adventitious roots, inhibition of lateral buds, stimulation of the cambium, and failure of senescent leaves

to abciss (LOCKE, RIKER and DUGGAR 1938) are elicited when rapidly growing crown gall tumors are induced in host plants such as the tomato.
Slowly growing relatively benign crown gall tumors do not, on the other
hand, elicit these responses from a host such as the tomato.

The auxin-independence of crown gall tumor tissue is further evidenced
by the fact that all bacteria-free tumor tissues thus far isolated from
many different plant species do not require an exogenous source of auxin
for their continued growth in culture as do most normal plant cells.
Rapidly growing cultured crown gall tumor tissue is, furthermore, generally
far more hydrated than is homologous normal tissue. Since water uptake
is known to be an auxin-dependent process, the tumor tissue must have
a very effective endogenous auxin-synthesizing system. When, moreover.
a culture medium upon which fully altered crown gall tumor tissue is
planted is supplemented with an auxin, that auxin does not stimulate the
growth of most tumor cells and may exert a pronounced inhibitory effect
on their development. The failure of even small concentrations of auxin
to stimulate growth of fully altered tumor cells has been interpreted to
mean that amounts of this substance needed for optimal growth are
synthesized by the tumor cell itself.

Direct assays for auxin present in cultured tumor and normal tissues
of *Scorzonera* by KULESCHA and GAUTHERET (1948) have shown that tumor
tissues contain considerably more free auxin than do the normal tissues.
The extraction of the auxin was shown, moreover, by KULESCHA (1954) to be
instantaneous from crown gall tumor tissue but to be extracted slowly and
progressively from the normal tissue.

Findings such as these led to the rather obvious suggestion that crown
gall tumor cells have acquired, as a result of their alteration, a capacity to
synthesize in greater than regulatory amounts a growth-promoting substance of the auxin type (BRAUN 1947 b; DE ROPP 1947 a; GAUTHERET 1947 a).

The fact that crown gall tumor cells are hyperauxinic does not necessarily mean that the tumor tissue produces auxin more rapidly than does
normal tissue for the tumors do not grow as rapidly as does, for example,
healthy meristematic tissue at the apex of the root or shoot. What it does
indicate is that the mechanism that regulates auxin levels in normal cells
fails to operate in the tumor. Attempts have been made by several investigators to learn why the crown gall tumor cell produces auxin in
greater than regulatory amounts. These studies, which are of essentially
two distinct types, are described below.

In 1947 TANG and BONNER isolated, and in 1953 GALSTON, BONNER and
BAKER described the components of an enzyme system present in certain
plant species that destroys indoleacetic acid by oxidation. GALSTON and
DAHLBERG (1954) carried this concept one step further and reported that
indoleacetic acid oxidase activity of etiolated pea tissue is directly related
to and is a measure of the physiological age of the tissue. Thus in normal
stem, root and bud tissue capable of rapid growth the activity of indoleacetic acid oxidase was found to be very low. Slower growing tissues

had somewhat higher oxidase levels, whereas extracts of mature tissue showed the highest observed activity of this enzyme. These findings suggest that ageing and maturation of plant cells are the result of the increased production of indoleacetic acid oxidase. These workers reported, further, that the increased production of this enzyme with increasing age of the cells was due to its adaptive formation in response to elevated auxin levels. This system for regulating auxin is almost certainly not the only one of its kind in plants and, indeed, its very existence as a functional system in the living plant cell has recently been questioned (Bonner 1957). During recent years considerable research has been directed toward the effect of certain unsaturated lactones including coumarin and its derivatives on the growth of various plants and plant organs. The discovery by Goodwin and Kavanagh (1948, 1949, 1952) and Goodwin and Pollock (1954) that scopoletin (6-methoxy-7-hydroxy coumarin) is distributed in the oat root in a marked morphological gradient raises the question as to whether this substance plays a regulatory role in the normal growth and differentation of the root.

Bitancourt (1949) applied the concepts and methods of Tang and Bonner (1947) to crown gall tumor and normal tissue of the same type and reported the presence of indoleacetic acid-inactivating enzymes in normal tissue but not in crown gall tumor tissue. According to Bitancourt, then, the high level of auxin found in the tumor cells results from a decreased auxin destruction by those cells. If this concept is correct it would offer a simple straightforward answer to the observed auxin autonomy in crown gall tumor cells. However, Platt (1954) has recently looked into this question in considerable detail. On the basis of this carefully conducted investigation Platt concluded that it is highly unlikely that a peroxidative system could be active in the control of growth by indoleacetic acid oxidation in the crown gall tumor and normal tissues used in his studies. If, as Platt's work suggests, normal tissues do not destroy auxin at a faster rate than do tumor tissues, the alternative explanation for auxin autonomy in crown gall tumor cells is that such cells acquire as a result of their alteration the capacity for an increased endogenous synthesis of that substance. Essentially two papers dealing with this aspect of the problem are worthy of note. Henderson and Bonner (1952), like Platt, found no detectable indoleacetic acid-inactivating enzymes in either tumor or normal tissue of the sunflower. They did, however, report the presence of an inhibitor in normal sunflower tissue which prevented the conversion of tryptophan to indoleacetic acid. This inhibitor, which was absent in tumor cells, while effective in the presence of low concentrations of tryptophan was much less effective at higher tryptophan levels. According to this view, then, normal tissue, but not tumor tissue, contains an inhibitor of the enzyme that converts tryptophan to indoleacetic acid. Working along similar lines Nitsch (1956) found that extracts of normal tissues, which are not auxin-independent, do not contain appreciable quantities of indole-acetic acid, whereas this substance could be detected in extracts of the tumor tissue. Neutral auxins such as indole-3-acetonitrile and ethyl indole acetate were found in both normal and tumor extracts. On the basis of these results NITSCH postulated that perhaps the normal tissues manufacture appreciable amounts of a substance like indole-3-acetonitrile but are unable to convert it to indole-3-acetic acid, whereas the auxin-independent tumor tissues can. BITANCOURT (1954) and SCHWARZ, DIERBERGER and BITANCOURT (1955) have reported that, from a qualitative aspect at least, auxin metabolism is similar in both normal and tumor tissue.

There seems to be no question about the fact that auxin plays an important etiological role in the continued abnormal growth of the crown gall tumor cell. It is not yet clear, however, just how the tumor-inducing principle associated with this disease causes crown gall tumor cells to become autonomous for auxin. It is not known whether there is an increased endogenous synthesis of auxin by the tumor cell or whether some subtle and as yet uncharacterized auxin-inactivating system normally concerned with the regulation of growth is destroyed as a result of the action of the tumor-inducing principle.

2. Substances limiting for cell division

In 1951 STEWARD and CAPLIN reported a finding of considerable significance. These workers found that potato parenchyma cells could be stimulated to rapid growth and division as the result of the synergistic effect of 2,4-dichlorophenoxyacetic acid and coconut milk. Both substances were required for the development of a profuse growth by the tissue. In the same year NEWCOMB (1951) reported that JABLONSKI and SKOOG had observed that tobacco pith parenchyma cells enlarged greatly in the presence of indole-3-acetic acid without, however, dividing. It was later found (JABLONSKI and SKOOG 1954) that pith cells could be made to grow and divide actively if they were treated with both indole-3-acetic acid and coconut milk. SHANTZ and STEWARD (1952) have shown that there are at least three distinct biologically active factors in coconut milk capable of promoting cell division. One of these has now been identified as 1,3-diphenylurea (SHANTZ and STEWARD 1955).

Factors possessing comparable biological activity have been demonstrated in extracts of a number of plant tissues (SKOOG 1954 a, b; STEWARD and SHANTZ 1956). MILLER et al. (1955 a, b) showed, moreover, that 6-furfurylaminopurine is highly active in promoting cell division in normal tobacco pith tissue. This substance, which has now been synthesized and given the trivial name kinetin, was originally isolated from heated solutions of herring sperm desoxyribonucleic acid. This compound does not appear to occur naturally in plants. It is clear from the results reported in the literature thus far that the factors that are effective in promoting cell division represent not a single unique class of substances but they are, instead, a group of chemically unrelated compounds that perform, perhaps by different means, a rather specific function.

Extracts of crown gall tumor tissue, like coconut milk, have been shown to be a good source of the factors that are limiting for cell division in tobacco pith tissue (BRAUN and NÄF 1954; STEWARD, CAPLIN and SHANTZ 1955). The studies of BRAUN and NÄF (1954) have shown, furthermore, that

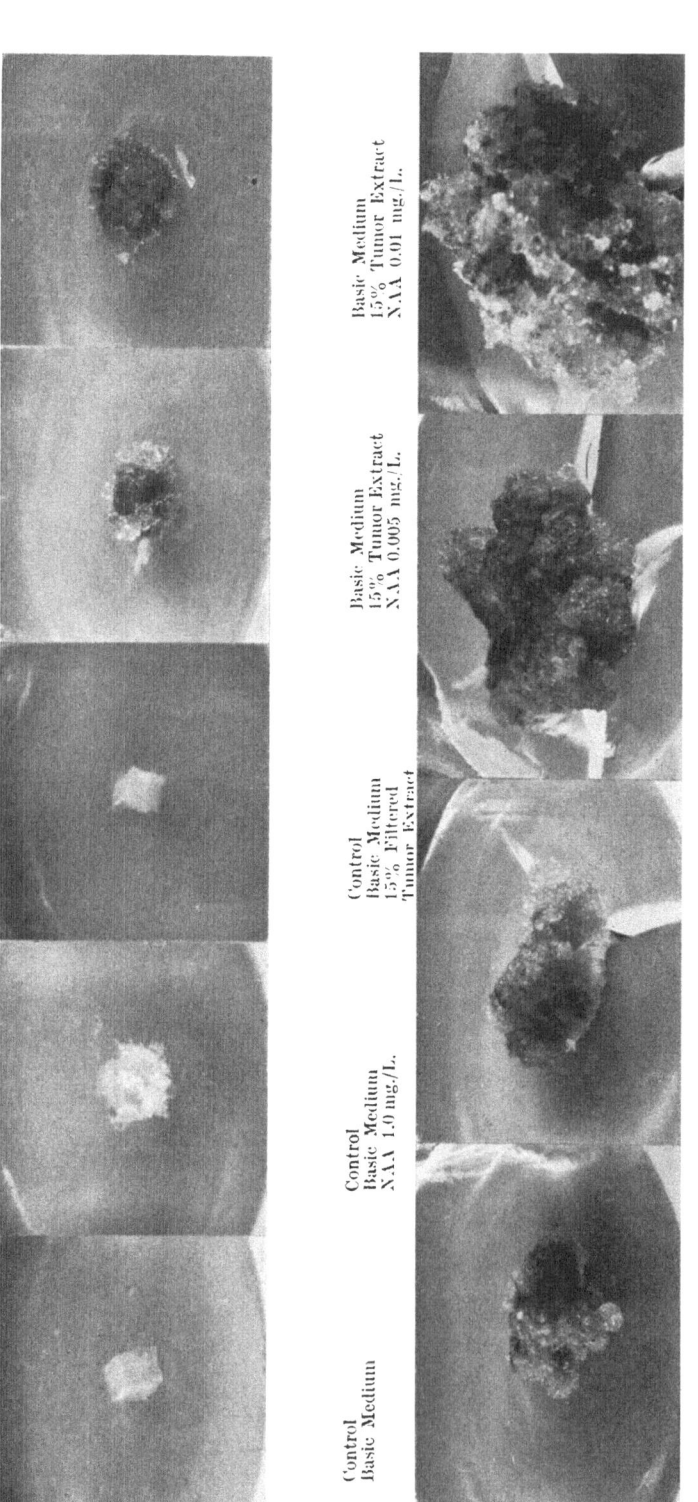

Fig. 4. Normal tobacco pith tissue planted on White's basic culture medium and treated as indicated above. Note that a direct relationship exists between the concentration of naphthalene acetic acid (NAA) in a medium containing a constant volume of tumor extract and hence cell division factor, and the amount and rate of growth that occurred in the normal tobacco pith tissue. The increase in volume of the tissue shown in the naphthalene acetic acid control is due entirely to cell enlargement. In those instances in which both naphthalene acetic acid and tumor extract are added to the medium, growth is the result of both cell enlargement and cell division.

(Photographs by J. A. Carlile.)

a correlation exists between the concentration of auxin in a culture medium containing a constant volume of tumor extract and hence cell division factor(s) and the amount and rate of growth accompanied by cell division that occurs in normal tobacco pith tissue fragments (Fig. 4). When high levels of auxin (0.5–1.0 mg./liter) were incorporated in a culture medium

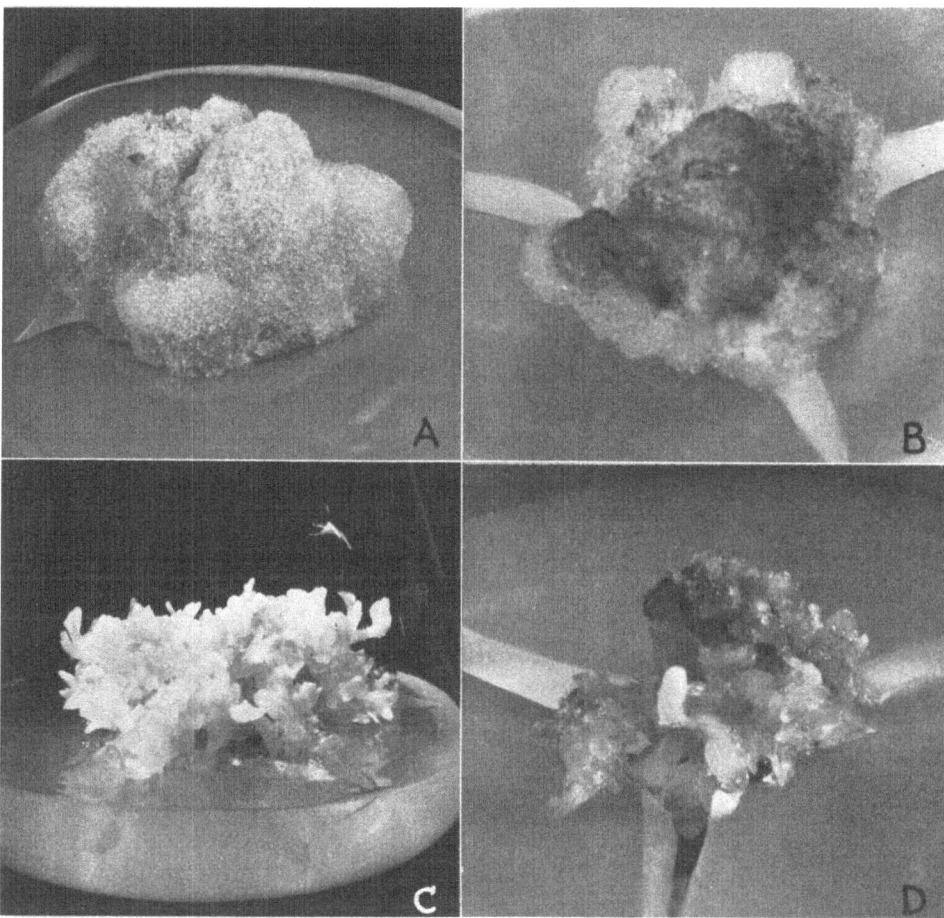

Fig. 5. A comparison of growth patterns of crown gall tumor tissue and of normal tobacco pith tissue stimulated by supplementing the basic culture medium with naphthalene acetic acid and an extract of crown gall tumors. A, Tobacco crown gall tumor tissue of the unorganized type grown on WHITE's basic medium. B, Normal tobacco pith parenchyma tissue grown on the basic medium supplemented with 0.5 mgs./liter naphthalene acetic acid and 15 per cent tumor extract. C, Crown gall teratoma tissue of tobacco on WHITE's basic medium. D, Normal tobacco pith tissue on the basic medium supplemented with 0.01 mgs./liter naphthalene acetic acid and 15 per cent tumor extract. Note abnormal but organized structures on the surfaces of C and D.
(Photographs by J. A. CARLILE.)

containing tumor extract, the normal pith fragments grew rapidly and in a completely unorganized manner. The growth pattern of the normal tobacco pith tissue in this instance showed a striking histological and morphological resemblance to crown gall tumor tissue of the unorganized type grown on the basic medium (Fig. 5, A, B). When the concentration of auxin was reduced in the tumor-extract-containing medium to a level equal to 0.01 mg./liter, the normal pith grew rather slowly and in an unorganized manner for about 4 weeks. Thereafter numerous more or less organized structures covered the surfaces of some of the normal tobacco pith tissue fragments. Such tissue showed a resemblance to crown gall teratoma tissue of tobacco (Fig. 5, C, D). It has thus been possible to reproduce under controlled conditions growth patterns that resemble, superficially at least, the morphologically distinct types of crown gall tumors that have thus far been described as occurring on tobacco. The artificially stimulated tissues were, however, self-limiting and when the externally supplied stimuli were removed, their growth promptly stopped. They were growth-substance stimulated hyperplasias. Crown gall tumor tissue is, on the other hand, autonomous and is itself capable of synthesizing growth factors needed for its continued abnormal growth.

It seems quite clear from studies of the type reported above that at least two growth factors, one of which is an auxin which is concerned with cell enlargement and the other which promotes cell division when used in association with an auxin, are essential if growth accompanied by cell division is to occur in tobacco pith cells. The pith cells have apparently lost as a result of their maturation the capacity to produce physiologically effective concentrations of these two growth substances. Since both growth-substance synthesizing systems appear to be solidly blocked in pith cells, these specialized cell types were used as test objects to gain insight into the nature of the cellular alteration in crown gall (Braun 1956 a). It was demonstrated that, as a result of the transformation of normal tobacco pith cells to crown gall tumor cells, the affected cells acquire the capacity to synthesize these two growth substances in greater than regulatory amounts and hence acquire an autonomy for both an auxin and the factor normally limiting for cell division in pith cells. This was evidenced by the fact that, whereas normal mature pith cells were unable to produce physiologically effective concentrations of either growth factor prior to their alternation to tumor cells, following transformation both factors were produced. If both growth-substance synthesizing systems had not been permanently activated, continued growth and cell division and hence tumor formation would not have occurred in the test system used in these experiments. Thus an essential difference between the normal tobacco pith cell and the crown gall tumor cell appears to be concerned at a physiological level with the permanent unblocking of two growth-substance synthesizing systems. This results in the continued production by the affected cell of greater than regulatory amounts of these biologically active substances which have been shown to be essential if growth accompanied by cell division is to occur in tobacco pith tissue. That both types of growth substances were actively synthesized by tumor tissue during growth was demonstrated, further, by grafting a fragment of bacteria-free crown gall tumor tissue of tobacco on a small block of normal tobacco pith tissue in culture. When the graft was successful, a pronounced stimulation of the pith cells involving both cell enlargement and cell division was observed within a 3-week period.

3. Other factors required for rapid autonomous growth of crown gall tumor cells

The permanent activation of the two growth-substance synthesizing systems described above with the resulting production of greater than regulatory amounts of the cell enlargement and cell division factors by the tumor cell would appear in itself to be adequate to account for the continued abnormal proliferation of the crown gall tumor cell. Recent studies

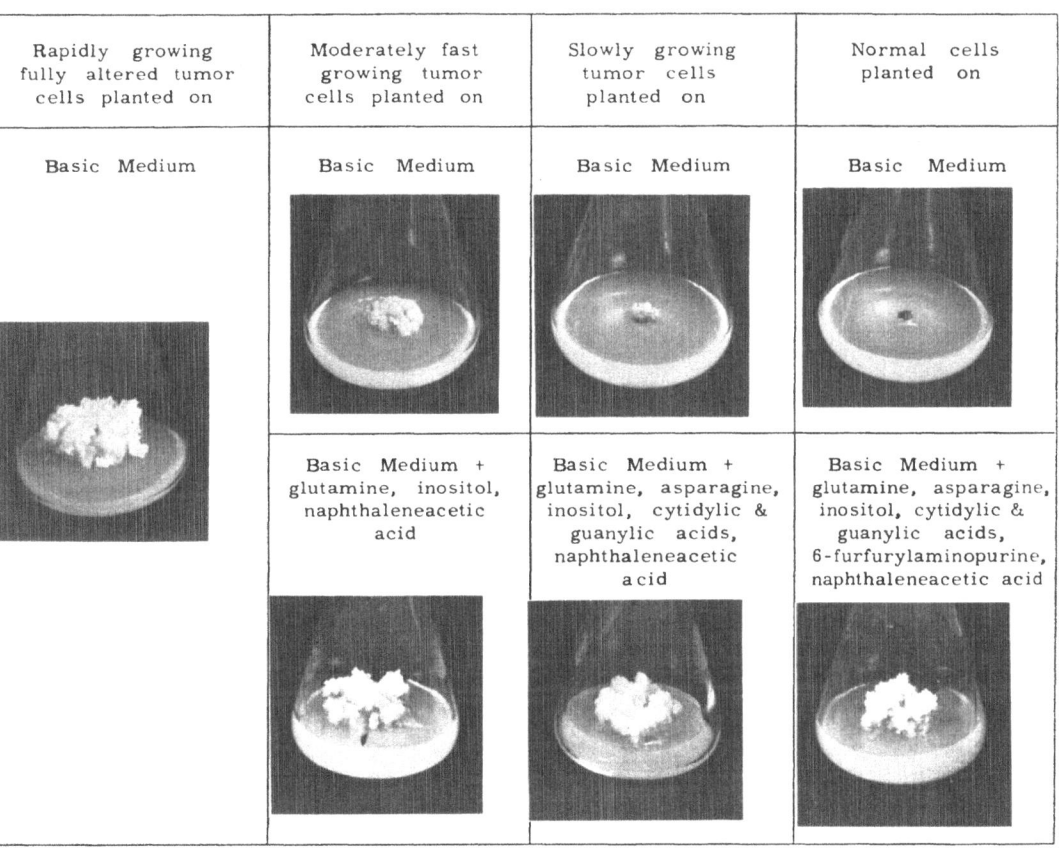

Rapidly growing fully altered tumor cells planted on	Moderately fast growing tumor cells planted on	Slowly growing tumor cells planted on	Normal cells planted on
Basic Medium	Basic Medium	Basic Medium	Basic Medium
	Basic Medium + glutamine, inositol, naphthaleneacetic acid	Basic Medium + glutamine, asparagine, inositol, cytidylic & guanylic acids, naphthaleneacetic acid	Basic Medium + glutamine, asparagine, inositol, cytidylic & guanylic acids, 6-furfurylaminopurine, naphthaleneacetic acid

Fig. 6. Relative rates of growth of three clones of crown gall tumor tissue that show different degrees of neoplastic change, planted on WHITE's basic medium. (*Left*), Fully altered, rapidly growing tumor cells. (*Upper left*), Moderately fast-growing tumor cells. (*Upper center*), Very slowly growing tumor cells. (*Upper right*), Normal cells of the type from which the tumor cells were derived. While the three clones of the tumor cells grow continuously although at different rates on the basic culture medium, normal cells of this type do not grow on that medium. Lower pictures and legends show minimal nutritional supplements needed by the three types of tissues to achieve a growth rate comparable to that of the fully altered tumor cell.

(Photographs by J. A. CARLILE)

have shown, however, that the permanent activation of additional growth-substance systems accompany the transformation of normal cells to tumor cells in crown gall.

It has been found that the alteration of normal plant cells to tumor cells takes place gradually (BRAUN 1943, 1947 a, 1951 b). This process can, moreover, be very accurately controlled experimentally (BRAUN and MANDLE

1948). When the tumor-inducing principle responsible for the alteration of normal cells to crown gall tumor cells acts on susceptible host cells for a 4-day period, rapidly growing potentially malignant tumors result. When, on the other hand, this carcinogenic principle is allowed to act for only 34 hours before being inactivated selectively by a thermal treatment, small slowly growing relatively benign tumors are initiated. A 50–60 hour exposure of susceptible cells to this principle results in tumors that grow in a host at a moderately fast rate. A similar type of tumor can be obtained by allowing the tumor-inducing principle elaborated by a moderately virulent strain of the inciting bacteria to act on host cells throughout a 4-day period.

Sterile tissues isolated from three types of tumors similar to those described above and planted on White's basic medium have maintained their characteristic growth patterns in culture for periods up to 10 years (Fig. 6). Normal tissue of the type from which the tumor tissue was derived does not grow on the basic culture medium.

Since these three tumor tissues, which showed varying degrees of neo-plastic change, were derived from the same plant species they were admirably suited for a further characterization of the factors that are limiting for rapid autonomous growth. In these studies the rapidly growing fully altered tumor cell was used as the standard. This cell type can synthesize in optimal or near optimal amounts all of the growth factors needed for its continued rapid proliferation from the inorganic salts and sucrose present in the basic culture medium. The moderately fast growing cell was found to require in addition to a cell enlargement factor (naph-thalene acetic acid) the vitamin meso inositol as well as glutamine to achieve a growth rate comparable to that of the fully altered tumor cell. It is interesting to note that a factor such as the naturally occurring equiva-lent of 6-furfurylaminopurine is not one of the growth factors that is limiting for rapid growth in this type of tumor cell, although this sub-stance is required for the growth of normal tissue. The enriched but chemically defined basic culture medium does not contain this factor and yet these tumor cells proliferate very actively. These results suggest that the cell-division-factor synthesizing system is very lightly blocked in normal cells of this type and that a low degree of cellular alteration that results in the partial activation of certain other growth-substance syn-thesizing systems is sufficient to permit the full activation of this system.

The nutritional requirements necessary to increase the growth level of very slowly growing cells altered in a 34-hour period to that of fully altered tumor cells are more exacting than are those required to accomplish this end when a moderately fast growing clone of tumor cells is used as the test object. In addition to an auxin, meso inositol and glutamine, asparagine, guanylic and cytidylic acids are required to accomplish this end. These studies demonstrate that, while normal cells altered to tumor cells in 4-day period are themselves capable of synthesizing optimal amounts of all growth factors needed for their continued rapid proliferation, cells transformed by the bacteria in shorter periods have had their requirements in terms of rapid growth only partially satisfied for several essential growth factors. It therefore follows that as the crown gall tumor cell becomes more autonomous its growth requirements in terms of externally supplied growth factors become less exacting. It thus appears that the transition from a slowly growing to a rapidly growing tumor cell is a gradual one involving quantitative rather than qualitative changes.

The results described thus far indicate, then, that several quite distinct metabolic systems are permanently activated as a result of the alteration of normal plant cells to tumor cells. This leads to the production by the affected cell of greater than regulatory amounts of substances concerned with growth and cell division. Since several distinct growth-substance synthesizing systems appear to be activated gradually during the transition from normal to fully altered tumor cells, the results suggest that there is some as yet uncharacterized master reaction that is specifically but gradually unblocked by the tumor-inducing principle and which, once activated, not only accomplishes the unblocking of several other growth-substance synthesizing systems but also determines the rate at which the entire series of metabolic events concerned with growth and cell division proceed.

Normal cells of the type from which the three clones of tumor cells described above were derived do not grow on the basic medium, nor do they grow on this medium when it is supplemented with substances that have been shown to be essential for the rapid growth of tumor cells possessing a low grade of neoplastic change. While the difference between the slowly and rapidly growing tumor cell is of quantitative nature since both can grow, although at different rates, continuously on the basic medium, the difference between the normal cell and the tumor cell is qualitative. The requirements for the continued rapid growth of the normal cells have now been characterized. While such a growth substance as 6-furfuryl-aminopurine is not essential for the continued rapid growth of any of the three clones of tumor tissue studied, such a factor is absolutely essential for the continued growth in culture of the normal cells. The absolute requirement of 6-furfurylaminopurine, or the naturally occurring equivalent of that substance, for the continued growth of normal cells in culture represents, then, a basic difference between a normal cell and a tumor cell. Normal cells, unlike the tumor cells, also possess an absolute requirement for an auxin for their continued proliferation in culture. The addition of 6-furfurylaminopurine and naphthalene acetic acid to the basic medium permits the very slow and limited growth of normal cells. However, only if the basic medium is supplemented with glutamine, asparagine, inositol, guanylic and cytidylic acids in addition to the auxin and 6-furfurylamino-purine, do the normal cells achieve a growth rate comparable to that of the fully altered, rapidly growing type of tumor cell. These studies, which are summarized in Fig. 6, indicate then that it is possible for a cell to acquire the capacity for autonomous growth as a result of the permanent activation of several well defined growth-substance synthesizing systems the products of which are concerned specifically with growth accompanied by cell division. These systems are precisely regulated in normal plant cells.

The problems now confronting the student of the crown gall disease are essentially twofold: *1*) to learn how the several distinct growth-substance synthesizing systems become permanently activated as a result of the action of the tumor-inducing principle, and *2*) to translate the information that is now available at a physiological level into biochemical terms. A knowledge of the precise nature of the tumor-inducing principle would doubtless assist in learning how this factor exerts its carcinogenic effects on a plant cell.

F. The tumor-inducing principle

The characterization of the tumor-inducing principle responsible for the alteration of normal plant cells to tumor cells has, during the past 40 years, presented a very real challenge to students of the crown gall disease. In those many studies in which success has been claimed it has invariably been shown by subsequent work that the reported results were in error due either to faulty technique or, in some instances, to an incorrect interpretation of the experimental findings. The early efforts in this direction have been reviewed in detail by Riker and Berge (1935) and Braun (1954 a) and will not be considered here. Only the more recent studies, some of which may eventually be shown to be significant, will be critically examined.

Although little, aside from its biological activity, is known concerning the nature of the tumor-inducing principle, it seems reasonable to suppose that in origin it may fall into one of the following five categories (Braun 1947 a). This principle may be either *1*) a metabolic product of the crown gall bacterium, 2) a normal host constituent that is converted by the bacteria into a carcinogenic principle, 3) a chemical fraction of the bacterial cell that is capable of initiating, as in the case of the transforming substance (desoxyribonucleic acid) of the pneumococci, a specific alteration with a resultant continued and, in this instance, abnormal development of those cells, 4) a virus associated with and transmitted by the crown gall bacteria, 5) the crown gall bacteria themselves, which enter the wounded cells and become so altered in their morphology and physiology as not to be demonstrable by either isolation or staining procedures.

Because of the precise control that can be exercised over the transformation process (Braun and Mandle 1948) and because of the striking temperature-dependence of the inception phase, it became possible to conduct inactivation studies over a critical temperature range of 23⁰ C. to 30⁰ C. (Braun 1950). The results of this study showed that measurable inactivation was limited to less than 2⁰ C. Despite this narrow range it was nevertheless possible to establish reaction rates and hence the energy of activation could be studied provied one assumption was made. If it was assumed that inactivation was due to thermal destruction, then the activation energy for this destruction could be computed according to the Arrhenius equation. When the energy of activation was evaluated from the experimental results obtained, very high values of more than 80,000 cal./mole were found. Since reactions of this order of magnitude are characteristic of protein denaturation, the results suggested that either the tumor-inducing principle itself or something intimately associated with the inability of this principle to initiate tumors at the higher temperatures may be a factor of complex structure. Needless to say, the validity of this conclusion rests on the validity of the assumption made in this study.

More recently studies have been carried out by KLEIN (1952, 1953, 1954; KLEIN and KNUPP 1957; KLEIN, RASCH and SWIFT 1953), the results of which have been interpreted to indicate that a specific polymer of desoxyribonucleic acid (DNA) may, in fact, be the tumor-inducing principle. It was reported (1952, 1953) that in tomato plants the levels of DNA in the host tissue increased 200 per cent of control values within 24 hours after inoculation with virulent crown gall bacteria and remained at that level for an additional 24 hours, after which time it dropped abruptly, again reaching control levels 72 hours post-inoculation. In contrast to the results obtained when a virulent culture of crown gall bacteria was used as the source of inoculum, tissues treated with avirulent cultures showed no DNA peak. When, moreover, a tomato plant inoculated with the virulent strain was held at 30° C., the observed DNA level was reported to be less than half of that found at 25° C. 24 hours after inoculation. At the higher temperature the curve broke sharply, reaching control levels after 48 hours, suggesting, according to KLEIN, a heat-induced depolymerization of the DNA.

In broad bean (*Vicia faba*) a 33 per cent increase in DNA was observed which did not relate to an increase in the bacterial population or to an increase in the number of host cells. Quantitative cytochemical methods failed to demonstrate any increase in FEULGEN-positive material of host nuclei or cytoplasm (KLEIN, RASCH and SWIFT 1953).

Much of the above information can be integrated very well into what is known about the time during which the tumor-inducing principle is active in transforming normal cells to tumor cells in certain host plants. However, from the evidence presented thus far it is difficult to judge whether such a reported rise in DNA is the cause or the effect of the transformation process. Furthermore, it can be demonstrated experimentally that in the tomato, in contrast with the response obtained in certain other hosts, smaller, far more slowly growing tumors are initiated in the first 48 hours following inoculation than are induced in the same number of hours between the second and fourth days or the third and fifth days after inoculation. In the first instance the DNA level at the point of inoculation would, according to KLEIN, be maximal, while in the last instance it would again have declined to the level found in the uninoculated control plants. It is possible, although this is not suggested by KLEIN's data, that a second DNA peak develops as normal cells are altered to tumor cells in the latter two instances. If such a second peak cannot be demonstrated by critical experimentation, it would certainly indicate that Klein's hypothesis, on the basis of the data presented, is wrong.

There are other aspects which should be considered. If the tumor-inducing principle is a chemical factor such as the DNA postulated by Klein, then one may consider it in terms of the following three possibilities listed earlier: 1) as part of the bacterial cell itself, 2) as a metabolic product of the bacterium, 3) as a normal host constituent converted by the bacteria into a carcinogenic principle.

The first possibility, that the observed rise in DNA reflects DNA which is actually part of the bacterial thallus, is least likely. Klein (1954) negates this possibility and points to the fact that there is no corresponding rise in RNA as would be expected if bacterial multiplication were involved. Other information also confirms this. The phosphorus content of the crown gall bacteria as calculated from P^{32}-uptake studies is of the order of 2.5×10^{-9} μg. per bacterium (Stonier 1956 a). In tomato, of the order of 700 μg. DNA-phosphorus (DNAP) are produced per gram (dry weight) of tissue (Klein 1952). On the assumption that 20 per cent of the bacterial phosphorus is in the (tumor-inducing) DNA fraction, such a quantity would represent 1.4×10^{12} bacteria. (Should the percentage of such DNA-phosphorus be less, the number of bacteria involved would be correspondingly more.) This means that 24 hours post-inoculation a slice of tomato stem one millimeter thick (dry weight approximately 1.5 mg.) would contain a minimum of 2.1 billion bacteria, enough to solidly occupy about 1.6 cubic millimeters (average bacterial volume $= 0.783$ $μ^3$: Stonier 1956 a). This is in contrast to common experience.

Therefore the second possibility, that a much smaller number of bacteria synthesize and release large amounts of nucleic acid, is more probable. If this is true, one might expect that the phosphorus metabolism of virulent bacteria differs from that of avirulent crown gall bacteria and that this would be reflected in the amount of phosphorus released by P^{32}-labeled bacteria. This is not the case (Stonier 1955, and unpublished material).

The third possibility, that a normal host constituent becomes converted by the bacteria into a carcinogenic agent, presents two alternatives. The host provides the DNA precursor which is assimilated by the bacteria, altered, and then released. Or, the host provides the DNA precursor(s) and an alteration is accomplished extracellularly by means of an enzyme or other product released by the bacteria. The first of these two hypotheses involves a special case of the tumor-inducing principle's being a bacterial metabolite, and, as will be shown below, is also in conflict with P^{32}-release studies. The second category implies that the enzyme or other bacterial product as well as the polymerized DNA may induce tumors in the absence of bacteria.

Klein's case would be greatly strengthened, of course, if the tumor-inducing principle could be demonstrated *in vitro*, and Klein (1954) did subsequently report that he was able to induce tumors with sterile filtrates of bacterial cultures. Since the biological activity was reported to be lost following heating of the solution and as the result of the addition of protamine sulfate, it was believed that the active substance was DNA.

In these studies virulent crown gall bacteria were grown in filter-sterilized juice, obtained from conditioned tomato plants, for 3 days at 24⁰ C. with constant shaking. Following incubation, the preparation was centrifuged at 8000 r. p. m. for 10 minutes to remove the bulk of the bacteria. The supernatant fluid was filtered through a Selas 02 filter and introduced through tubes into wounded decapitated tomato plants (personal communication to one of us, A. C. B.). The percentage of successful induction of tumors by this technique was of the order of 10 to 30 per cent. The use of juice obtained from previously heavily wounded plants was essential to the process; ordinary plant juice failed to exhibit any activity.

Subsequent studies (STONIER 1953, and unpublished work) again failed to demonstrate significant differences in the release of phosphorus by virulent P³²-labeled bacteria when grown in juice obtained from conditioned plants as compared with juice obtained from ordinary plants. More important, studies on the effectiveness of Selas filters by BRAUN (unpublished work) indicated that sterile preparations cannot be obtained with the use of a Selas 02 filter, and hence studies based on their use must be disregarded.

In dealing with filtration studies a number of factors must be kept in mind. Filters have a saturation point. There is a finite limit to the amount of contaminated fluid that may be passed through them before leakage occurs. In the case of *Agrobacterium tumefaciens*, this saturation point is very quickly reached for most of the standard filters used in the preparation of sterile media. Studies by one of us (A. C. B., unpublished results) on the Selas filters, for example, indicate that not only is the 02 porosity size useless, but so are the 03 and 04 sizes except where small volumes of liquid are concerned. Better results may be obtained if solutions are filtered in series. It must be remembered that a single crown gall bacterium is sufficient to induce a tumor (HILDEBRAND 1942). The tumor will arise once bacteria find a proper environment and start multiplying. The chances for this occurrence depend on the size of the inoculum and the size of the wound. Furthermore, HILDEBRAND (1942) found that about one-third of the small overgrowths initiated with small inocula proved to be sterile. In view of these findings it becomes imperative to use large samples of a fluid when testing for its sterility.

It is seldom that negative results are reported in the literature and so, many, probably hundreds of experiments dealing with filtered extracts of tumors, bacterial cultures, etc., have remained unrecorded simply because no success was achieved. An exception is the recent report by BENDER and BRUCKER (1956) dealing with extensive attempts to obtain bacteria-free tumor induction. The few times these workers obtained tumors in their experimental plants were when the bacteria had somehow contaminated

the plant tissue.   In this connection one might also recall the paper of
Rosen (1926) who reported that small forms of the organism pass through
Berkefeld filters.

Very recently Klein and Knupp (1957) modified considerably the earlier
procedure.   In these experiments, in contrast to the previous studies, all
preparations were sterilized by successive filtration through Selas 015, 02,
and 03 filters, and discs of carrot phloem tissue were used as the test object.
The carrot discs were isolated sterilly and planted for 48 hours on White's
basic medium containing naphthalene acetic acid at a concentration of
1 part per 10 billion parts of medium.   The discs were then placed in juice
from strong reactor carrots for 24 hours.   This treatment was followed by
three 24-hour exposures of the discs to a filter-sterilized water extract of
malted barley supplemented with glucose, in which the bacteria had grown
for 24 hours.   The tissue was then treated with 1.5×10^{-6} M indoleacetic
acid at pH 7.0 to accomplish "promotion."   The carrot juice, the malt
extract, and the bacterial culture were all dialyzed in 0.001 M phosphate
buffer, pH 7.0, to remove toxic components.   Each step was critical to the
success of the experiment.   Small isolated growths were obtained which,
however, according to Klein and Knupp (1957), never reached a size comparable to those initiated by the bacteria.   According to the authors it was
not possible for technical reasons to graft these growths into healthy hosts
to demonstrate their neoplastic properties.   The paper fails, moreover, to
mention certain types of controls that would appear essential if conclusions are to be drawn from this study.

Although the tumor-inducing principle may eventually be shown to be
a specific polymer of DNA, the information available at this time does not
warrant such a conclusion.

A characterization of the tumor-inducing principle would doubtless be
facilitated if evidence could be brought to bear on the question as to
whether this principle is itself a self-duplicating entity that is the permanent agent of change and is responsible for the continued abnormal
proliferation of the crown gall tumor cell or whether this factor, like
many of the carcinogenic substances and radiant energy, evokes a heritable
change in the affected cell and is, therefore, concerned only with the inception of the tumor.   The products of the biochemical lesion resulting
from the action of the tumor-inducing principle on a susceptible cell would,
in the latter instance, be the continuing cause of tumor cell growth.   Such
a heritable change could conceivably be carried either by the nucleus or
by the cytoplasm.

The possibility that the bacteria themselves act as the tumor-inducing
principle by becoming intimately associated with the host cells was recently
reëxamined by tracing the fate of P^{32}-labeled bacteria in the host by
means of radioautography.   The radioactivity remained localized in the
intercellular spaces, except for cells whose walls had been ruptured, or
xylem elements.   It was concluded that the bacteria exert their effect across
the host cell membrane (Stonier 1956 b).

The possibility of a virus being transmitted by the bacteria as the causal agent in the crown gall disease has received attention in recent years. If a virus is concerned etiologically in crown gall, it is certainly not one that is readily transmissible either mechanically or with the use of special techniques suitable for the transmission of those plant viruses that are not readily transmissible.

The virus hypothesis was first seriously advanced by DE ROPP (1947 b) after he had reported the *in vitro* development of "secondary tumors" in normal sunflower stocks above which were grafted fragments of bacteria-free tumor tissue scions. This worker interpreted these findings as pointing to the presence of a transmissible agent in crown gall tumor tissue that incited the initiation of new tumorous growths in the tissue of the healthy stock. If this interpretation is correct, then crown gall is almost certainly a virus disease. Similar results were obtained by McEWEN (1952) as well as by CAMUS and GAUTHERET (1948). The latter used Jerusalem artichoke, *Helianthus tuberosus*, as the test object. DE ROPP's pictures indicate that the "secondary tumors" are quite distinct morphologically and histologically from the original tumor fragment. However, as discussed in the section dealing with the histology of crown gall tumors, interpretations other than one invoking the concept of secondary tumor induction are equally plausible. It may be, for example, that both tumors are actually derived from cells of the same original tumor fragment, the difference in structure and growth rate being simply a reflection of the position in the stem that the original tumor cells occupy.

The virus hypothesis received further impetus as a result of a report by CAMUS, WILDMAN and BONNER (1951) which was discussed by GAUTHERET (1952). The results of this study indicate that crown gall tumor tissue contains a new high molecular weight protein that constitutes about 20 per cent of the total protein. This new protein was non-infectious and was reported to be absent in normal and habituated tissue of the same kind. Although the original studies were reported in 1951, a full account of the work has not yet appeared. The etiological implications of this finding may, of course, be very great.

BEARDSLEY (1955) has recently shown that the highly virulent B 6 strain of the crown gall bacterium carries a lysogenic phage. An isolate has been obtained from this strain of bacteria that is free of the phage and is itself susceptible to the phage. This bacterial isolate was found to be as virulent in inducing tumors in plants as was the virus-carrying parent strain (personal communication). This particular bacterial virus is not, therefore, involved etiologically in the crown gall disease.

MANIGAULT, COMANDON and SLIZEWICZ (1956) have published a paper, the results of which, if confirmed, can only be interpreted as implicating etiologically an entity of very large size in crown gall. The experimental procedure used by these workers was simple and straightforward. In these studies the bacteria were grown in juice obtained from conditioned tomato plants. Following a 48-hour incubation period, the juice was centrifuged at a low speed and filtered to free it of bacteria. The filters used were the Seitz AW (coarse) on top of an EKI (fine) filter (personal communication). The tumor-inducing principle was then found in *1*) the filtered extract, in *2*) the pellet of a first high (100,000 × g) speed centrifugation, in *3*) both the pellet and the supernatant of a subsequent low (6,600 × g) speed centrifugation, and in *4*) the pellet of the second high speed centrifugation. Two items in the data throw doubt upon the conclusion that the tumor-inducing principle is a virus. Firstly, the sizes of the overgrowths pictured, with the exception of Fig. 6, are so small that it is very questionable whether they are tumors. Secondly, if it is granted that they represent tumors and not simply wound response, most if not all of the activity seems to sediment at 6,600 × g (see their Figs. 3 and 4). No true virus would sediment, except perhaps by coprecipitation, at this speed. It would take a particle the size of an ordinary bacterium to sediment at that slow speed. The authors of this review have now carried out the procedure outlined above on two separate occasions and on a rather large scale. The filters used by Manigault et al. have proved effective. However, inoculations of *Datura* and *Pelargonium* plants, the hosts used by Manigault et al., and of *Kalanchoe* and tomato plants with filter-sterilized tomato extract upon which crown gall bacteria had grown as well as with pellets obtained by centrifuging such extracts at 6,600 × g and 100,000 × g have never yielded tumorous overgrowths in our hands. Considerable differences were found, however, in the amount of normal wound healing that occurred following the various treatments as well as in the untreated control plants. Additional studies along these lines should be carried out.

The results of one additional series of experiments may serve to help characterize the tumor-inducing principle. In studying the origin of the crown gall tumor cell Braun (1951 a, 1954 b) attempted to distinguish between somatic mutation at the genic level and the possibility that a self-duplicating autonomous or semi-autonomous entity had assumed control of the cells and was responsible for the continued abnormal proliferation of crown gall tumor cells. It is well known in biology (Kunkel 1924; Sonne-born 1946) that certain self-duplicating cytoplasmic entities can be eliminated from cells under conditions that favor the increased multiplication of such cells over and above the multiplication rate of the self-duplicating cytoplasmic particle. It is known, furthermore, that normal meristematic cells found at the apices of rapidly growing roots and buds divide at far faster rates than do most tumor cells. It was therefore hypothesized that, if crown gall tumor cells could be made to organize buds and if those tumor buds could be forced into rapid growth, recovery of the tumor cells might be accomplished provided the factor responsible for the continued abnormal cellular proliferation in crown gall was subject to the effects of dilution in very rapidly dividing cells.

Crown gall tumor cells of the commonly found type are characterized by their rapid and uncoördinated powers of proliferation, their very limited powers of differentiation, and their complete lack of a capacity to organize. Since such tumor cells have never shown the slightest tendency

to become less autonomous, they have generally been regarded as being
permanently altered cells and therefore unsuited for investigations on
recovery. It was found, however, that when pluripotent cells of certain
plant species were altered to tumor cells by a moderately virulent strain
of the crown gall bacterium, there was produced in place of the characteristic
crown gall tumor a complex overgrowth or teratoma. These teratomata
were composed of chaotic assemblies of tissues and organs that showed
varying degrees of morphological development. Sterile tissue isolated from
such abnormal but organized structures grew profusely, as did tumor
tissue of the commonly found type, on a culture medium that does not
support the continued growth of normal cells. The teratoma-derived tissue
retained indefinitely, moreover, the capacity to organize abnormal leaves
and buds. The striking organizational capacity of such teratoma-derived
tissues seems to reflect the inherent regenerative potentialities of the pluripotent tumor cells and does not in itself appear to affect directly the
recovery of those cells. The bacteria-free teratomata provided the necessary experimental tool for studies on the recovery of crown gall tumor
cells. When bacteria-free teratoma tissues were grafted to the cut stem
tops of suitable plants, abnormal tumor shoots developed from the tumor
buds present in the tissue. Such abnormal shoots were forced into rapid
growth by a series of graftings to healthy plants. When these tumor cells
were forced to very rapid growth in this manner, they gradually recovered
and became normal in every respect. The fact that it is possible to cause
a reversion of such tumor cells to completely normal cells as a result of
rapid but organized growth suggests that crown gall tumor cells contain
within themselves all of the cellular factors, both genetic and non-
genetic, that were present in the normal cells from which the tumor
cells were originally derived. In addition, however, the tumor cell
appears to have acquired as a result of bacterial action an additional
factor which, if the interpretation given the results is correct, is subject to
the effects of dilution and is therefore lost in very rapidly dividing cells.
Whether this factor is a virus transmitted by the bacteria or whether the
tumor-inducing principle somehow specifically modifies or selects certain
classes of normal self-duplicating cytoplasmic entities, or whether in fact
some entirely different mechanism is involved, remains to be determined.
With the recent development of methods by MUIR, HILDEBRANDT and RIKER
(1954) for growing single plant cells, the problem of the recovery of the
crown-gall tumor cell should be reinvestigated.

Discussion of the crown gall disease

Although the tumor-inducing principle responsible for the alteration
of normal plant cells to tumor cells in the crown gall disease remains
uncharacterized, insight into the manner in which a plant cell responds to
the action of that principle has been gained. The transition from a normal
cell to a fully altered tumor cell has been found to be a gradual process.
Cells possessing varying degrees of neoplastic change can, therefore, be

obtained at will and grown in culture under precisely defined environmental conditions. Results obtained with the use of tissue culture methods have permitted not only a characterization of the factors essential for rapid autonomous growth, but they have given an indication of significant differences that exist between a normal cell and a crown gall tumor cell. The capacity for autonomous growth in the crown gall tumor cell appears to be the direct result of the permanent activation of a series of growth-substance synthesizing systems, the products of which are concerned specifically with growth accompanied by cell division. The permanent unblocking of these growth-substance systems, which takes place gradually during the transformation process, makes available to the tumor cell greater than regulatory amounts of these essential growth substances. The resulting hormonal imbalance could and probably does account for the continued abnormal poliferation of the tumor cell. Fully altered, rapidly growing tumor cells are themselves capable of synthesizing, from mineral salts and a carbon source, optimal or near optimal amounts of all growth factors required for their continued abnormal growth. Cells showing lower grades of neoplastic change have had their requirements in terms of rapid growth only partially satisfied for several essential growth factors. The different degrees of autonomy shown by crown gall tumor cells appear, therefore, to be a reflection of the availability to the tumor cell of several essential growth-promoting substances that are synthesized by the cells themselves as a result of their alteration. The growth-substance imbalance found in crown gall tumor cells appears not only to account for the continued abnormal growth of such cells but also appears to play an essential role in determining the morphological, histological, and cytological characteristics of the tumor tissue.

The student of plant growth has at his disposal chemically characterized substances with which he can control, under precisely defined experimental conditions, such fundamental growth processes as cell enlargement and cell division. By varying the concentrations of a growth factor concerned with cell enlargement and that limiting for cell division in a chemically defined culture medium, growth patterns are produced in certain normal tissues that bear a striking resemblance to those described for crown gall tumor tissues. When, for example, the ratio of the cell enlargement factor to a cell division factor is great, a rapid completely unorganized growth results. Such normal self-limiting growths are composed of hyperplastic and hypertrophied cells. Multinucleate giant cells as well as other nuclear abnormalities commonly found in crown gall tumor tissue are present. Reversing the ratio of these two growth substances results in a more slowly growing compact tissue composed mostly of hyperplastic cells. Under these conditions organization of cells into tissues and organs may occur. Such growths bear a superficial resemblance to crown gall teratomata. It is thus possible to reproduce under fully controlled conditions the several growth patterns that have been described for crown gall tumor tissue.

The fundamental question as to how the tumor-inducing principle activates growth-substance synthesizing systems within a cell remains unanswered. This principle may exert its effect by destroying in whole or in part the cellular system that normally regulates the growth-substance level within a cell or, conversely, it may somehow cause the affected cells to synthesize excessive amounts of several essential growth substances. In either case, once the cellular system is modified, the alteration is perpetuated from one generation to the next. Such a change might conceivably be carried either in the nucleus or in the cytoplasm. Studies on the recovery of the crown gall tumor cell favor the latter possibility. A virus transmitted by the inciting bacterium of the crown gall disease could account for all of the reported findings. Experimental evidence implicating a virus is lacking, however.

Of interest in this connection is the relationship that exists between the normal wound healing reaction and tumor inception. This relationship, which is presented schematically in Fig. 7, shows that tumors are not initiated in the first 24 hours or subsequent to the fifth day after inoculation of the bacteria into the plant by wounding. Very small slowly growing tumors are initiated either in the first 34 hours or subsequent to 90 hours after inoculation. A minimum period of 60 hours is required for the initiation of large rapidly growing tumors. It is at this period in the wound-healing cycle that cells show the greatest metabolic activity. These results can be accounted for entirely on the basis of the need for conditioned host cells if the cellular transformation is to occur. Conditioning results from the stimulus of a wound and renders the host cells susceptible to alteration. This process takes place gradually; it reaches a maximum between the second and third days after a wound is made and declines again as wound healing progresses toward completion. Histological studies have shown that it is just before or during the earliest stages of active wound healing that normal cells are altered to tumor cells of the most actively growing type. It is clear, therefore, that both the tumor-inducing principle and those as yet uncharacterized factors that make a host cell vulnerable to transformation are critical in determining whether the cellular alteration will occur. The degree of neoplastic change achieved in a cell appears, moreover, to be dependent upon both factors.

The conditioning effect, which renders the host cells susceptible to alteration, may involve permeability changes which enable the tumor-inducing principle to penetrate conditioned cells. In the absence of evidence of such changes, these results are understandable if it is assumed that a normal cellular component, of as yet unknown nature, is elaborated in gradually increasing amounts during the early stages of wound healing, reaching a maximum concentration between the second and third days after a wound is made. At that time the cells in the region of the wound begin to divide, and since this component is assumed no longer to be elaborated after cell division begins, its concentration would decrease in proportion to the number of divisions that the cells had undergone during the healing of the wound. This hypothetical component could conceivably be any one of the several classes of particulate cytoplasmic entities capable of replication. Such entities are commonly found to be present in greatest

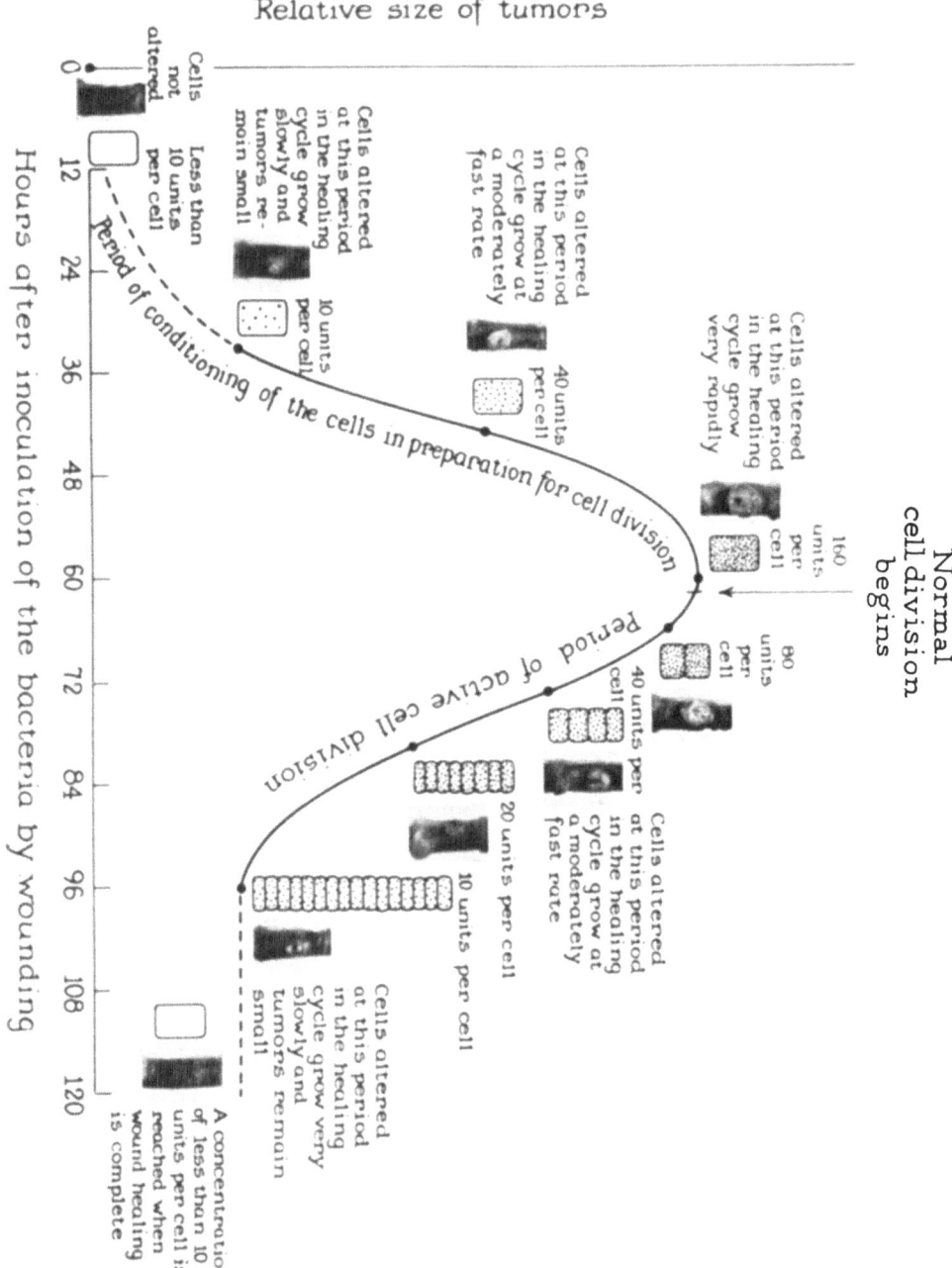

Fig. 7. The relation of normal wound healing to tumor inception in crown gall. Conditioning of the host cells resulting from the stimulus of a wound is essential if the cellular alteration is to occur. The conditioning process, which is schematically represented here, takes place gradually, reaching a maximum between the second and third day after wounding and declining again as normal cell division begins and healing progresses toward completion. The size of tumors produced can be correlated with the period in the normal healing cycle in which the cellular transformation is accomplished.

(Photograph by J. A. Carlile.)

numbers in cells that show high metabolic activity. Their replication and
persistence appear, moreover, to be under gene control and to be carefully
regulated by the cell.

Essentially three mechanisms can be visualized by which the tumor-
inducing principle in crown gall might affect the particulate population of
the cytoplasm. 1) It might favor the selection of one segment of the population by specifically inhibiting other types. This would result in a selective
replication with the resulting dominance of one type over another and could
lead to serious metabolic disturbances in the cell. 2) The tumor-inducing
principle might act in such a way as to fix permanently the number of
hypothetical components that are present in the cell at the time of transformation as a result of conditioning. Thus cells transformed 34 or 96 hours
after wounding would contain fewer such entities than would cells altered
60 hours after wounding. According to this interpretation the number of
such particles would not only be fixed but would be maintained at a constant
level by the tumor cell as it proliferates. The time required for conditioning would, in this instance, represent the period of time necessary for
the host cells to elaborate the specific cytoplasmic component. 3) The
third mechanism by which the tumor-inducing principle might affect the
particulate population of the cytoplasm is by combining with or otherwise modifying existing elements in such a way as to form new self-
perpetuating combinations with these components. Thus, instead of being
diluted out as is the normal component during the course of wound
healing, the new combination would continue to develop within the cell.
To account for the observed results it must again be assumed, as in the
previous instance, that the maximum concentration of this new entity in
the tumor cell is regulated by the cell and does not commonly reach a level
greater than that present in the cell at the time that the cellular transformation occurs. This newly formed component would be capable of
directing the activities of enzyme systems in much the same manner as do
the carcinogenic viruses.

That not only the concentration of the hypothetical cytoplasmic entity
but also the concentration of the tumor-inducing principle is critical in
determining the degree of the cellular alteration is suggested by the following observation. When a highly virulent strain of the crown gall
bacterium is used to transform normal plant cells to tumor cells, tumors
of the largest most rapidly growing type are initiated 60 to 72 hours after
wounding. Slowly growing tumors are induced when the transformation
occurs at a somewhat earlier or later period in the healing cycle. In this
instance the degree of cellular alteration reached would appear to be concerned with the concentration in the host cells of the hypothetical normal
component. On the other hand, cells altered to tumor cells by an attenuated
culture during the optimal period for transformation in the healing cycle
grow very slowly and behave as do cells altered by the virulent strain
during the early and late stages of wound healing. The limiting factor in
this instance would appear to be the relatively smaller amounts of tumor-
inducing principle available for combination with the normal component.

According to this concept the degree of cellular alteration reached would be dependent both on the amount of tumor-inducing principle available and on the concentration in the activated host cells of the hypothetical normal component. While this concept remains a working hypothesis, it does appear to explain quite satisfactorily the experimental findings.

Much of this speculation could be circumvented if it were possible to isolate, characterize, and trace the fate of the tumor-inducing principle in the host cell. Once this is accomplished it should be possible to gain insight into its mode of action.

Habituation

Permanent modifications in plant cells that are quite similar in certain respects to the cellular transformations in the crown gall disease have been recorded. Gautheret (1946) found, for example, that when normal cells are isolated from a plant such as *Scorzonera,* the cells possess an absolute requirement for auxin if they are to grow in culture. After such tissues had undergone numerous transfers in a culture medium containing an auxin in the form of indoleacetic or naphthalene acetic acid, sectors occasionally developed from the tissue fragments that possessed a capacity to grow indefinitely on an auxin-free medium. Such cells, like crown gall tumor cells, may become completely insensitive to auxin. Gautheret has given the name *"accoutumance à l'auxine"* or *"anergie à l'auxine"* to this phenomenon. American authors have commonly used the term *habituation.*

Shortly after the original observation by Gautheret, Morel (1947) found that this phenomenon also occurs in tissue cultures of the European grape, tobacco, mallow, and the Virginia creeper. Somewhat later, Kandler (1952) and, independently, Henderson (1954) reported it for the sunflower, while Czosnowski (1952 b) found it to occur in tissue cultures of the carrot. Habituated or anergized tissue, in addition to being auxin-independent, is commonly friable in texture, possesses only a slight capacity for differentiation, and never produces roots. Such tissue has, according to Kulescha and Gautheret (1948) and Kulescha (1952), a moderately high level of free auxin. Czosnowski (1952 a) has reported that habituated tissue of the European grape differs biochemically from normal tissue of the same species in many respects. Earlier, Lee (1952) had made a comparative study of certain nitrogenous constituents present in crown gall, habituated, and normal tissue of the European grape. No qualitative difference in the amino acid composition of these tissues was found. The crown gall tissue had the highest concentrations of total and soluble nitrogen, while the normal tissue had the smallest amounts. Habituated tissue had intermediate levels. Gautheret considers this phenomenon to be neither a mutation nor a selection but rather a kind of enzymatic adaptation leading to a profound modification in the chemistry of the cell. Unlike the cellular transformation in crown gall which can be very accurately controlled, habituation seems to occur only sporadically and cannot as yet be induced at will under controlled conditions. Of particular interest is the

reported capacity of certain of the auxin-independent tissues to form
tumors when grafted into healthy plants. LIMASSET and GAUTHERET (1950)
found, for example, that large overgrowths resulted when fragments of
habituated tissue of tobacco were grafted into tobacco plants. Less specta-
cular results were obtained (BRAUN and MOREL 1950; HENDERSON 1954) when
fragments of habituated tissues of other plant species were grafted into
their respective hosts.

It has been shown (MOREL 1947; GAUTHERET 1955) that there are degrees
of habituation, just as there are degrees of cellular transformation in the
crown gall disease. Of interest is GAUTHERET's (1955) suggestion that plant
tissues may become habituated to growth factors other than auxin.

Virus Tumors

Introduction

A second plant disease, wound tumor, which fulfills all the criteria of
a true neoplastic disease, is one caused by a typical virus. In nature this
virus, *Aureogenus magnivena* Black, is insect-transmitted. It may also be
transmitted by grafting and by mechanical means if high enough virus
concentrations are used. The virus is one of a group of viruses that multiplies in its animal as well as in its plant host. It has been isolated from
both hosts, purified, and observed with the electron microscope. On a large
variety of plants the signs produced by the virus involve morphogenetic
disturbances; in some plants such as sweet clover (*Melilotus alba* Desr.)
and sorrel (*Rumex acetosa* L.) the response to infection involves the production of large tumors which show a capacity for unlimited, disorganized
growth both *in vivo* and *in vitro*. For earlier reviews the reader is referred
to BLACK (1949, 1952, 1954).

The tumor-inciting virus was first isolated from the agallian leafhopper,
Agalliopsis novella (Say), by BLACK (1944), who found that young crimson
clover (*Trifolium incarnatum* L.) plants developed irregularly enlarged
leaf veins when exposed to virus-infected insects. Two other species,
Agallia constricta Van Duzee and *Agalliopsis quadripunctata* (Provancher),
were also found to act as vectors. A subsequent survey of potential plant
hosts indicated that at least 43 species in 20 families produced symptoms
typical of the disease, principally irregular enlargements of the veins in
leaves and woody tumors on the roots. Other symptoms observed included leaf curling and distortion, leafy outgrowths from the underside of
the veins, vein tumors, distortion of petioles, shortening of internodes,
thickening of stems, suppression of flowers, and dwarfing (BLACK 1945, 1952).

The inciting agent

The inciting virus is of general interest to biologists not only because of
its association with abnormal plant growth but also because it has the
ability to multiply in both plant and animal hosts. This was demonstrated

by feeding infected insects on Grimm alfalfa (*Medicago sativa* L.) which is immune from wound tumor virus. The insects were infected by injecting them mechanically with a clarified "brei" obtained from other infected insects (Maramorosch, Brakke and Black 1949). Black and Brakke (1952) showed that during 7 serial passages of this type a minimum dilution of 10^{-18} was achieved. However, the virus titer per insect remained constant from one generation to the next, and as much virus appeared to be present in the insects at the end of the experiment as in the beginning. Subsequently, Black (1953) was also able to show that a small percentage of females of *Agalliopsis novella* infected with the virus oviposited eggs which produced infective progeny. There may exist a correlation between multiplication of the virus in the insect host, insect transmission, and the fact that mechanical transmission is feasible only if high concentrations of the virus are used (Brakke, Vatter and Black 1954). After infection, the virus undergoes a minimum incubation period of 13 to 15 days at the optimum temperature before the insects are able to transmit it to healthy plants (Maramorosch 1950).

The infective agent has the following properties (Black, Maramorosch and Brakke 1950; Brakke, Maramorosch and Black 1953; Brakke, Vatter and Black 1954; Black 1955). It is filterable through filters retaining small bacteria and has a sedimentation constant (S_w^{20}) of about 600 Svedberg units. Electron microscope pictures of purified and dried preparations indicate a polyhedral particle approximately $80 \, m\mu$ in diameter. This is true whether the virus is obtained from infected plant or insect material. The virus particle is moderately stable. When suspended in neutral saline, it may withstand 50^0 C. for 10 minutes but is destroyed at 60^0 C. under the same conditions. Virus activity was demonstrated in desiccated material stored at 0^0 C. for one year.

In vitro studies

Once the virus has infected a plant tissue and initiated tumors, it remains closely associated with the tumorous host tissue. The tumors of sweet clover contain about one hundred times more virus than do the non-tumorous portions of the diseased plant (Brakke, Vatter and Black 1954). It is possible to culture tissue derived aseptically from tumors. Such tissue was isolated from sorrel tumors and found to contain virus after 14 months of culture. The tissue doubled in volume about every 3 weeks on the original medium. However, Burkholder and Nickell (1949) subsequently devised a more successful medium which permitted much more rapid growth. The medium consists of mineral salts, sucrose, and vitamin B_1 (Nickell 1952). The tumor tissue requires unusually high concentrations of phosphorus in the medium. Originally this was interpreted to reflect the multiplication of virus-like nucleoprotein in the infected tissues (Burkholder and Nickell 1949). Similarly, the observation that an RNA hydrolysate, and specifically uracil, is stimulatory would be in accord with this hypothesis. DNA hydrolysate, adenine, adenosine, adenylic acid, triazolo analogues of guanine, and 2,6-diaminopurine are all inhibitory. Guanine, xanthine, and hypoxanthine, as well as uracil, prove stimulatory (NICKELL, GREENFIELD and BURKHOLDER 1950; NICKELL 1954). Synthetic medium on which such tumor tissue had grown for some time was found to contain an α-amylase (BRAKKE and NICKELL 1951, 1955).

Tumor-initiating stimuli

In this, as in other neoplastic plant diseases, wounding plays an important part in the etiology of the disease. Tumors develop from accidental wounds, from points of stress, from points of emergence of lateral roots, or from emergence holes of a parasitic wasp, *Tetrastichus,* when the adult insect leaves the plant (TEITELBAUM and BLACK 1954). Needle punctures through infected stems start tumors, especially in young tissue. Once initiated, such tumors show an unlimited capacity for growth, both *in vivo* and *in vitro,* with the virus apparently continuing to infect the tissue. It is of interest to note here that when infected plants are treated with auxin, a profound increase in number and size of stem tumors occurs (BLACK and LEE 1957).

Morphological and genetic aspects

The variations of symptoms produced by the virus vary from barely detectable to those that are extremely pronounced. In *Portulaca oleracea* L., the tops of infected plants bear no symptoms of any sort; only the roots bear numerous small tumors. On the other hand, sorrel and sweet clover produce pronounced tumors, the latter on the stem. These two hosts show considerable individual variations in their response to infection. Investigations on sweet clover demonstrated that the heredity of the plant affects the number, size, distribution and morphology of the tumors (BLACK 1951). In some strains root tumors may be so numerous and large that they fuse together, while in other strains they may be so small and inconspicuous as to be easily overlooked. Stem reaction in some clones involves the formation of many large tumors, in other clones stem tumors rarely appear. Hereditary influences on the stem and root may act independently. For example, one clone produces large tumors on stems and roots, while another produces many large root tumors but only a few stem tumors. However, in general, it seems that conditions favoring vigorous plant growth favor a greater tumor response, and *vice versa.* Heredity influences the distribution of the tumors. One clonal reaction consists of a tendency to form stems swollen throughout a considerable part of their length. Similarly, the morphology of stem tumors appears to be genetically determined. One clone may produce elongated tumors with a rough surface, another, globose tumors with a smooth surface. BLACK has interpreted these observations in terms of many nuclear genes. However, no experimental evidence presented rules out the possibility that at least some of the factors might involve cytoplasmic inheritance.

That the genetic basis which causes a clone to be highly predisposed to tumor formation involves not merely a susceptibility to the virus but a tendency toward tumorous proliferation is indicated by the inbred B 21 line of sweet clover. This line, which appears to be inclined to tumor formation, produced 5 spontaneous tumors over a period of several years (Littau and Black 1952 b). That these did not represent accidental infections was demonstrated by the facts that no root tumors were observed on these plants which ordinarily would possess such root tumors, and that these plants showed no immunity to infection by wound tumor virus. The 5 spontaneous tumors had an identical and distinct histology. The authors compare line B 21 with strain C_3 Hb of mice which have lost the mammary carcinoma virus yet still show an inherent tendency to mammary carcinoma (Heston et al. 1950).

Histological and cytological aspects

The root tumors of sweet clover originate in the pericycle opposite the primary phloem and immediately adjacent to the endodermis (Kelly and Black 1949). In the early stages the tumors follow a definite and regular pattern of development. Abnormal phloem tissue differentiates first in the basal or innermost portion of the young tumors and is followed by the differentiation of the "apical" or outermost cells into abnormal xylem elements. These tracheids tend to be of irregular shapes and sizes and differentiate from the tip toward the base in finger-like processes. This cap of tracheids prevents further "apical" growth of the tumor. The meristematic cells between the xylem and phloem continue to grow and often push around the differentiated xylem "apex." Ultimately, the organization of the tumor deteriorates. The tumor contains groups of tracheids some of which are connected to other groups by ramifying tracheids; these groups are often surrounded by parenchymatous cells through which small branches of abnormal phloem are interspersed. It would appear from these studies that meristematic tumor cells are capable of becoming differentiated.

In extending these observations Lee (1953 a) reported that 85 per cent of the root tumors are initiated in the pericycle close to young lateral roots. Lee describes the formation of lateral roots as follows. In the zone of maturation the pericycle is a uniseriate layer located adjacent to the endodermis and encircling the central vascular system. Usually the stelar tissue consists of three primary phloem strands alternating with the triarch primary xylem, separated by one or two layers of cambium cells. The formation of lateral roots starts with periclinal divisions of a single pericyclic cell close to a protoxylem element and is followed by similar divisions of a plate of adjacent pericyclic cells. The cells of the endodermis are eventually stretched by the lateral root primordium and also undergo divisons. The cortical cells around the newly formed root are slightly crushed and pushed aside by the emerging root.

Lateral root formation in young roots occurs in the maturation zone at definite intervals. The earliest meristematic tumor cell can usually be detected next to the fourth or fifth incipient lateral root from the main root tip. At this stage the lateral root is emerging from the main root, and the meristematic tumor cells appear around the base of the lateral root.

in the pericycle, usually opposite the primary phloem but also frequently close to the protoxylem. The first division of the pericycle is, as in the case of the lateral root, periclinal. However, it is almost immediately followed by inclined divisions with cell walls forming in random directions. The meristematic cells of an incipient tumor undergo repeated division, the successively formed cells being of smaller and smaller size. Subsequently the cells of the tumor begin to differentiate, as described by KELLY and BLACK (1949).

As indicated, tumor formation usually occurs later than lateral root development. The growth of a lateral root sometimes continues after a tumor has formed at its base; often, however, it is interrupted and much retarded. Some such roots degenerate or become short stumps encircled by tumor growth. In seriously infected roots, chains of tumors, gradually increasing in size from less than 1 mm. to 1 cm. in diameter, can be found from the root tip upward. In advanced stages chains of small tumors sometimes occur on the main root between the bases of two lateral roots. These probably arise from virus-infected cells in the pericycle of the root. These tumors have a structure similar to that of the young tumors around the lateral root except that the various parts are better defined, as in stem tumors of the same plant. Sometimes tumors occur simultaneously in parts of the pericycle opposite the primary phloem, even without the presence of lateral roots. One of these usually grows more rapidly than the other two, although sometimes the three encircle part of the main root.

LEE (1955 a) also described the formation of virus tumors associated with nitrogen-fixing bacterial nodules. The situation is analogous to lateral root formation; the nitrogen-fixing nodules appear earlier on the roots than do the tumors. Shortly after the appearance of the nodules, tumors develop at their bases. The nodules, unlike the lateral roots, have no definite position relative to the normal vascular bundles of the root since the bacteria attack root hairs at random. Nevertheless, the incipient tumor cells are usually located in various portions of the pericycle opposite the primary phloem. A visible retardation and disintegration of many young bacterial nodules result; in only a few is the damage from such tumors slight or inapparent.

In the stems of sweet clover the tumors arise in the phloem. The first divisions usually involve phloem parenchyma or phloem fibers, occasionally the phloic procambium (KELLY and BLACK 1949; LEE 1955 b). Older tumors often exhibit a relatively high degree of organization with an anatomy similar to that of the normal stem. A region of meristematic tumor cells differentiates internally into abnormal xylem, externally into a zone of "pseudophloem." The whole tumor mass is surrounded by what appears to be a normal cortex and epidermis, the cortical cells having enlarged but slightly. The vascular tissues of the tumor may show a connection with those of the stele, and the meristematic tumor zone may be continuous with the stelar cambium.

Pronounced differentiation of scalariform tracheids has also been observed in tissue cultures of sorrel tumors (BURKHOLDER and NICKELL 1949).
In one case such cultured tissue had differentiated leaf-like organs about
4 years after isolation; the formation of apparently normal-looking roots
with root hairs, although a rare occurrence, does happen from time to time
(NICKELL 1954, 1955). Whether these structures arise from normal tissue
or from tumor tissue remains a matter of conjecture. It has not been
possible to culture these roots on media which support normal *Rumex* roots
in vitro. This has been construed as indicating that perhaps such roots
are not "normal" roots (NICKELL 1955). *In vivo*, virus-free cuttings may be
obtained from infected sweet clover plants and a tumor probably contains
some secondarily stimulated normal tissue. It is possible that such tissue,
subjected to continuous stimulation by the tumor cells, continues to proliferate along with these cells. Only single cell experiments will provide
unequivocal evidence as to whether or not tumorous tissue is capable of
organizing. Although in sweet clover the tumors usually originate in
parenchymatous tissue associated with the phloem, in other plant species
the tumors may originate in other tissues (KELLY and BLACK 1949). In
Rumex, in addition to large root tumors, small wart-like tumors originate
from the cork cambium. In *Hyoscyamus* and *Lychnis,* irregular overgrowths of the cambium may occur.

Of all the neoplastic plant diseases, the virus-induced wound tumor
disease is unique in so far as it is the only one in which it is possible to
recognize some cells as tumor cells on the basis of their cytological characteristics alone. In a number of plant species tumor cells frequently contain many spherical cytoplasmic inclusions which have been named
spherules (LITTAU and BLACK 1952 a). These spherules may be found in
virus tumors of *Anagallis linifolia* L., *Chrysanthemum leucanthemum* L.
var. *pinnatifidum* Lecoq. and Lamotte, *Hyoscyamus niger* L., *Lychnis alba*
Mill., *Melilotus alba* Desr., *Nierembergia frutescens* Dur., *Rumex acetosa*
L., and *Silene latifolia* (Mill.) Britten and Rendle. Bodies of similar appearance are not found in meristematic root tips of healthy *Rumex* nor,
with a few rare exceptions, in healthy tissues of other species. The spherules are associated with cells of tumors resulting from the virus infection
and are not found in adjacent non-tumorous cells of the same plant. The
spherules are spherical in shape, vary in size up to 4μ in diameter and
stain red with safranin, pyronin and acid fuchsin. They are Feulgen-
negative and do not stain with IKI solution or Sudan IV. A strong positive reaction for arginine may be obtained with the SAKAGUCHI reaction.
However, the MILLON reaction, the BERG ninhydrin test, and the xantho-
proteic test fail to provide evidence for protein: the spherules are not di-
gested by pepsin, trypsin, chymotrypsin or papain when tried singly. The
evidence for ribonucleic acid is equivocal. However, it is not possible to
differentiate these cytoplasmic (not vacuolar) inclusions from nucleoli by
staining techniques except by means of a special nitrous acid-ribonuclease-
azure A technique described by LITTAU and BLACK (1952 a). In areas of
tumor growth the spherules serve as a marker of the virus-infected tumor

cells. Hence it is possible to ascertain whether tumor cells as such may differentiate. The answer is positive. The differentiation of cells containing spherules into tracheids has been described by several workers (LITTAU and BLACK 1952 a; LEE 1955 b; LEE and BLACK 1955).

In addition to differentiating, the virus-infected tumor cell may be prevented from continuing its unrestrained proliferation in quite another manner. Recently LEE (1956) has depicted the following process in the developing tumor of *Capsella Bursa-pastoris* (L.). After several tumor cells have formed, the nuclei become invisible. Instead, granules of various sizes may be found dispersed throughout the protoplasm. Some of these granules stain like chromatin with safranin. The granules become more abundant in later stages, being more or less evenly dispersed throughout the cell although their size becomes variable (on rare occasions up to 20 μ in diameter). The apparently enucleated cells containing these inclusion bodies then develop into "hyperplastic tumor cells" by the frequent and uneven subdivision of the tumor cell into many smaller cells. In old tumors the repeated divisions of tumor cells without any corresponding increase in volume results in so many small cells that individual cell structures become obscured. However, it is apparent that the walls of these cells are unevenly thickened and incompletely developed. It would be of considerable interest to know whether, as LEE (1956) suggests, 1) the virus infection actually results in a fragmentation of the nucleus, and 2) whether the apparently irregular chromatin fragments are sufficient to permit cytokinesis. If the granules are nuclear fragments, then they are not similar to the spherules described by LITTAU and BLACK, a possibility which LEE mentions. Appropriate cytochemical studies would help to resolve this problem.

Discussion of virus tumors

Some of the observations described above warrant a certain amount of additional discussion. First, it must be recognized that usually at least two factors are essential for the initiation of tumors: a wound or some similar stimulus such as lateral root initiation or hormone application, and virus infection. The function of the wound is not known. The release of wound hormones might cause the host cell to become susceptible to the action of the virus (LEE 1955 b). It has been suggested (BRAKKE, VATTER and BLACK 1954) that perhaps the cells' reaction to a wound permits an increase in virus concentration and that the high concentration maintains the cell multiplication. Another possibility is that the net result of the various stimuli described above is to expose dividing cells to viral action. That is, as a result of the presence of the virus, once cells of an infected tissue begin to divide, they are no longer subject to those host morphogenetic restraints which normally return the cell to quiescence after wound healing has been carried to completion. Of pertinence at this point is the question raised by KELLY and BLACK (1949) as to why one cell in the pericycle of an infected sweet clover root develops into an organized lateral root, while other pericycle cells in fairly close proximity develop into dis-

organized proliferating tumor cells. There are several possibilities. In view
of the fact that lateral root initiation occurs earlier than tumor initiation,
it may be that the more embryonic tissue has such tight morphogenetic restraints placed upon it that the infection does not lead to its disorganization. Or it may be that, since the lateral roots initially involve the
xylary pericycle and the tumors involve the phloic pericycle, these differences reflect the divergencies in morphogenetic response at the two sites.
A third possibility rests upon the assumption that the virus is only able
to infect, or infect heavily, certain cell types. It is known that certain
leafhopper-transmitted viruses are associated primarily, if not solely, with
the phloem tissue of the plant host. In sweet clover, lateral roots usually
start with cell divisions of the pericycle close to the protoxylem (LEE 1955 a).
In contrast, root tumors usually start in the pericycle opposite the primary
phloem (KELLY and BLACK 1949; LEE 1955 a). It may be that at that stage
of root maturation which permits lateral root initiation, *usually* only that
portion of the pericycle closest to the phloem is infected. Therefore not
until lateral root formation reaches such a stage as to involve tissue closer
to the phloem does a root tumor start, because not until then does the cell
division stimulus associated with lateral root formation reach virus-
infected cells. The same is true for wound tumor initiation at the base of
developing nitrogen-fixing root nodules. Similarly, in the stem of sweet
clover it is the phloem and pericycle in which growth is initiated. It must
be remembered that the pericycle is very closely related to phloem both
in terms of its anatomical proximity and its histogenic development
(ESAU 1953). In the petioles of crimson clover, *Trifolium incarnatum*, only
the phloic procambium, not the xylary procambium, gives rise to meristematic tumor cells (LEE and BLACK 1955). Again, in the root tumors of
Capsella Bursa-pastoris the tumors originate in the pericycle close to the
phloem (LEE 1956). It may be that the high virus content of sweet clover
tumors (BLACK 1955) is due to the large amount of "pseudo-phloem" contained by such tumors (LEE 1955 b). Thus, the type and amount of tissue
susceptible to virus multiplication is likely to be one of the factors which
would determine both the origin and the subsequent morphological features of the tumors. A hypothetical gradation may be considered with
Medicago sativa L. at one extreme in which no tissue whatsoever permits
virus multiplication and therefore no tumors develop, to *Melilotus alba* in
which the virus multiplies, perhaps, only in the phloem and pericycle, to
Rumex acetosa where other tissues, such as the cork-cambium, may also
become infected and contribute to tumor formation, to the other extreme
which involves a host, as yet not discovered, in which all tissues would be
susceptible to virus multiplication and consequent tumor formation. In
other words, one may consider that in the case of wound tumor, because
only a limited tissue is capable of becoming tumorous (because only a limited tissue is infected with virus), a more organized histological appearance
results. This is in contrast to crown gall in which, if wounding is sufficiently severe, a general transformation of all potentially parenchymatous
tissue takes place.

If the origin and subsequent morphological development of a tumor depend upon what tissues were initially infected at the time of tumor initiation, then they are equally dependent upon the initial size and shape of the initiating wound or other stimulus (hormone application). Here again the phloem may play a significant role in so far as it is known to be the chief transport system of plant hormones.

Once the tumor has been initiated, a third set of factors comes into play. These involve the capacity of the tumor cells to stimulate surrounding cells into proliferation. Tissue culture studies have indicated that the tumors are auxin-independent. It is likely that the tumor tissue produces an excess of hormones which will stimulate parenchymatous cells into proliferation. Histological observations tend to bear this out. The high degree of organization exhibited by *Melilotus alba* stem tumors initiated by pin-pricks probably reflects the fact that the cambium and cortex are secondarily stimulated to divide in a more or less regular fashion (*e. g.* Fig. 11, Kelly and Black 1949; Fig. 10, Lee 1955 b). Actually it is possible that tumor-produced hormones may play an even more drastic role than to induce proliferation of ground meristems. Figure 5 of Lee and Black (1955) suggests that perhaps specialized tissue such as the pericycle fibers redifferentiate and then contribute to tumor formation. Whether this is a primary or a secondary effect remains to be seen.

If one of the factors which determines virus tumor morphology involves the capacity of normal cells to behave like tumor cells as a result of secondary stimulation, the reverse must also be considered as an important factor. That is, the degree to which the tumor cell may be induced to behave as a normal cell and to differentiate. For although at the histological level there is much disorientation in the tumorous tissue, meristematic tumor cells appear capable of differentiation. This statement is based on two types of observations: the position of the differentiated elements which strongly suggests their tumorogenic origin, and the differentiation into tracheids of cells containing large numbers of spherules. Since the spherules appear to be a sign of virus infection in the tumor cells and are not present in normal meristematic cells, the conclusion is inescapable that tumor cells may differentiate. Differentiation appears to be limited, however, since vessels, or sieve tubes, are never observed in such tumor tissue.

If by definition a cancerous cell is one that shows a propensity toward a continued and unrestrained proliferation in the organism in which it is developing, and a tendency to be less differentiated than the normal cell from which it was derived, then the implication of the above studies is clear. The virus-induced neoplastic state is not necessarily irrevocable: it may be reversible. What specific factors suppress the neoplastic state is not known. In the case of Lee's observations on root tumors of *Capsella Bursa-pastoris*, it seems reasonable that if the virus or some other factor causes the nucleus to fragment, the cell will not be able to continue its proliferation. The arguments which indicate that an excess of hormones

5*

of the auxin type may be instrumental in the differentiation of tumor tracheids have been presented in the section dealing with the histology of crown gall. What favors the differentiation of phloem-like elements from tumor cells is not known.

The morphological features of a wound tumor depend, then, upon at least four factors: 1) the type of tissue infected with the virus; 2) the geometric pattern of the infected cells that are subjected to an initiating wound stimulus; 3) the types of normal tissues induced to divide along with the tumor tissue as a result of the growth-stimulating substances produced by the tumor; and 4) the extent to which the neoplastic state is suppressed, causing the tumor cell as well as the normal cells to undergo irreversible differentiation.

Genetic Tumors

Introduction

The production of spontaneous malformations is a regular occurrence in certain plants. No external agents, for example infectious parasites, are associated with these phenomena and in that sense the overgrowths may be considered to arise spontaneously. It is the genetic constitution of the plant which appears to be the primary cause. The most thoroughly investigated example involves the spontaneous formation of tumors on a number of *Nicotiana* hybrids. When appropriate *Nicotiana* species are crossed, tumors form in 100 per cent of the hybrid offspring. These tumors apparently arise in mature tissue which has been stimulated into cell division by natural or pathological processes, and once initiated, the tumors retain the capacity for unlimited disorganized growth both *in vivo* and *in vitro*.

Only relatively few cases of what appear to be genetic tumors have actually been investigated. Littau and Black (1952 b), as discussed in relation to virus tumors, have reported on the occasional formation of spontaneous tumors in their B 21 line of sweet clover, *Melilotus alba*. Kostoff (1933, p. 441) has stated that in certain *Crepis* and other species hybrids the spontaneous appearance of fasciations or tumorous proliferations has been observed. Unfortunately, Kostoff gives no references. Jensen (1918) has reported that multiple nodosities appear regularly at the roots of the hybrids of *Brassica campestris* × *B. napus*. However, as Jensen pointed out, these growths cannot be considered to be truly tumorous since they are self-limiting and fail to continue their growth upon transplantation. Satina, Rappaport and Blakeslee (1950) have described the occurrence of ovular tumors in interspecific hybrids of *Datura*.

In contrast to these limited observations, the tumors which arise regularly in certain interspecific hybrids within the genus *Nicotiana* have been intensively studied. In 1930 Kostoff reported that nine interspecific hybrids of *Nicotiana* produced spontaneous tumors. Since that time another forty or so tumor-producing combinations have been obtained, about half of them by KEHR (for a listing of almost all known tumor-producing hybrids to 1953, see KEHR and SMITH 1954).

Attempts to transmit a causative agent by grafting or to demonstrate microbial agents concerned with the etiology of the tumors have consistently failed (KOSTOFF 1930; KEHR and SMITH 1954; KUNKEL, personal communication). The evidence indicates that the pathological condition is genetically inherent in the plant and needs only some stimulus, which does not affect other plants in the same way, to bring out this undefined inherent property. Wounding is such a stimulus. In some of the hybrids it is possible to initiate a tumor at almost any time by making a wound in the vicinity of a vascular bundle. Normally, tumors on the stem form at the site of leaf abscission. On mature leaves, wounding by insects or mechanical damage leads to tumor formation. Tumors are found more commonly on the roots than on stems or leaves. A number of hybrid combinations produce tumors on roots only. In those instances in which tumors develop on all parts of the plant, root tumor formation often precedes formation of tumors on the shoot. This increased susceptibility to tumor formation on the part of the root may reflect several factors. Lateral root formation may represent a stimulus similar to wounding and may, therefore, play a role in tumor initiation analogous to that encountered in the virus tumors. This point cannot be cleared up, however, until the appropriate histogenetic studies have been performed. Chemical wounding may initiate tumors. KEHR and SMITH (1954) report that leaves accidentally sprayed with a mixture of turpentine, whiting, and white lead produced tumors at almost every spot where droplets of the spray mixture struck the leaf. These authors quote IZARD (1952), on the other hand, as being unable to induce tumors by treating the leaves with plant growth hormones of the auxin type and with several other substances. Feeding plants radioactive phosphorus, or irradiating them, hastens the onset of tumor formation and increases significantly the number of tumors that develops (SPARROW and GUNCKEL 1956; SPARROW et al. 1956). Until evidence to the contrary is brought forth, the action of these agents may be considered to cause nonspecific injury which in turn elicits a tumor-forming response. Recently, strong evidence has been provided which indicates that in these hybrids neither "spontaneous" tumor formation nor radiation-induced tumor formation involves a process of somatic mutation (H. H. SMITH 1958).

Morphology

The macroscopic appearance of these tumors has been described by many workers, some of whom have called attention to the fact that the outward appearance of the genetic tumors is similar to that of tumors initiated by crown gall bacteria. This is true, but needs to be qualified by adding that the genetic tumors resemble tumors initiated by moderately virulent strains of *Agrobacterium tumefaciens* on *Nicotiana langsdorffii* in that the stem tumors have a strong tendency to differentiate and organize. The genetic tumors seldom, if ever, reach the size or the state of disorganization of crown gall tumors initiated by highly virulent bacteria. The tumors are almost never directly fatal to the plant. They represent a burden to the plant, however, for it has been observed that meristematic activity virtually ceases on branches bearing tumors (Kehr and Smith 1954). These authors further report that under humid conditions the tumors are highly susceptible to rot organisms and are frequently decayed, while branches bearing them gradually become yellowish in color and eventually succumb. Kostoff (1930) believed the death of some plants to be due to root tumors.

Histology

Histologically, the first signs of stem tumor development appear in the tissue between epidermis and collenchyma (Brieger and Forster 1942). Certain dead or dying cells appear to provide the traumatic stimulus that causes adjacent cells to divide. In due time a whole region becomes activated and the cortex layer becomes converted to tumor tissue. The hyperplastic stimulus spreads to the medullary rays, the medulla, and finally the vascular cylinder is forced open, resulting in a mass of soft callus tissue consisting of hyperplastic and hypertrophic cells extending from the pith to the periphery of the tumor.

The more mature stem tumor of the *Nicotiana glauca* × *N. langsdorffii* hybrid has the following general histological characteristics (Kostoff 1930; Levine 1937; Kehr and Smith 1954). The proximal and basal portion of the tumor is continuous with the pith and consists of mature pith-like cells which stain lightly. Many of these homogeneous parenchymatous cells are filled with starch grains and differ in appearance from normal pith cells only in their position and the fact that the orientation of the long axis of the cell may be radial (toward the new tumor growth) instead of longitudinal. Vascularization of the tumors is relatively slight, but fibrovascular bundles tend to form an imperfect ring of tissue about a cone of "pith" and may connect with the vascular system of the plant. The peripheral portion of the tumor contains areas of embryonic growth which consist of small, deeply staining cells with large nuclei. These embryonic regions may give rise to distorted foliar and cauline elements. Delicate fibrovascular bundles arise from the plant and traverse the long axis of these abnormal leaf and stem elements; occasionally delicate venations may extend into the aberrant broad leaf-like embryonic structures. On the other hand, some of the partially differentiated shoots have no specialized conducting system at all.

It is in respect to vascularization that the hybrid tumors differ so greatly from crown gall or virus tumors. In crown gall tumors protoxylem cells and tracheids become differentiated early; later, bizarre knotting of xylem vessels or tracheids may be observed. In hybrid tumors not only is the vascularization of the overgrowths much lighter, but there is no evidence of any markedly irregular arrangement of these fundamental plant tissues. Levine has pointed out that the early death of some of the tumors suggests that vascularization is inadequate. He further considered the hybrid tumor with its calloid nature and tendency to form aberrant embryonic leaves and shoots as organoid; in contrast, he considered crown gall to be histoid.

Secondary growth often originates in the following fashion (Kostoff 1930). In regions where hypertrophy has followed the cessation of cell division, necrosis usually makes its appearance. Necrosis may occur in the midst of the tumor tissue or on the surface and is usually accompanied by autolysis. Necrosis of two or three layers of cells may be sufficient to reawaken cell division in the surrounding regions. Necrosis often affects as much as half of the tumor and shortly after the cessation of its advance there appear new outgrowths which cause the tumor to assume a different and irregular shape. As already indicated, these necrotic areas frequently become secondarily invaded by fungi and bacteria. However, Brieger and Forster (1942) have reported the presence of intracellular inclusion bodies found in the stems of *Nicotiana glauca* $\times$ *N. langsdorffii* during earliest phases of tumor formation, when the cells in the cortex become activated. These inclusions have an alveolar, highly refractive content, a colorless membrane, and resemble somewhat parasitic fungal cells such as those of the Chytridiales. The cells containing these inclusions do not participate in tumor formation. The authors stress the fact that they never have observed such inclusions in the parent species or in non-tumorous hybrids. Unfortunately, no extensive data are given. The authors refrain from interpreting the ascribed association of the inclusion bodies with tumorous hybrids. It may be that the hybrids studied were naturally susceptible to a particular parasite found in Brazil but that such a parasite has, otherwise, nothing to do with tumor formation. On the other hand, the parasite may exert a non-specific traumatic stimulus which initiates tumor formation. Finally, these inclusions might represent the product of an abnormal metabolism rather than a parasite. This last interpretation is unlikely since no such inclusions have been reported by any other workers.

Cytology

The tumor-producing cells have the following characteristics (Kostoff 1930). The intensely dividing cells are relatively small with densely staining cytoplasm. The plastids are small and reduced in number. Cellular polarity is lacking in the sense that cell division is disoriented. After the initial phase of division, the cell may undergo hypertrophy with numerous vacuoles of various sizes becoming evident. Whereas the dividing cell was small, the enlarged cell attains a size perhaps ten times that of the embryonic cells. Such enlarged cells are often rich in starch and tannins. The frequent occurrence of nuclear abnormalities is a characteristic of the tumor cells. The morphology of the chromosomes depends greatly upon the position of the cell in the tumor. The chromosomes of cells dividing in necrotic areas appear very much shortened and are often so short that they resemble those observed in generative cells. Occasionally individual chromosomes advance to the poles at mitosis; however, lagging on the spindle is more frequent. The chromosome separation may be a relatively slow process, the whole process of nuclear division being retarded. Chromatolysis may begin just after chromosome separation is effected with the result that nuclei are formed in very close apposition. Such nuclei are usually not separated by a cell wall and a binucleate cell results. In cases where the chromosome separation is very much delayed and chromatolysis coincides with separation, polyploid nuclei result. The nuclei in necrotic zones become hypertrophied and their shape becomes irregular. Autolysis first affects the nuclear membrane, then the nuclear content, finally the nucleolus. Kostoff (1930) points out that the polyploid and other abnormal nuclei are usually found in old tumors and near necrotic regions. He considered polyploidy to reflect an interference with the dynamics of nuclear division by substances diffusing from necrotic areas, and that nuclear abnormalities really represent a secondary phenomenon. Subsequently, however, Kostoff (1939) reported that abnormal mitoses in the root tips of "old" amphidiploid *Nicotiana glauca × N. langsdorffii* are a frequent occurrence, whereas no abnormalities are found in the root tips of *N. glauca* or of *N. langsdorffii*, nor in the autotetraploid of *N. glauca*. Kehr (1951; Kehr and Smith 1954) raised the interesting question that Kostoff may not have been dealing with a true amphidiploid, that it might, instead, have been a mixaploid. This interpretation gains support from Kostoff's report that the aforementioned tetraploid did not breed true. Kostoff (1930) also reported that the hybrid nuclei contain an increased number of nucleoli when compared with either parent. Kostoff, himself, discounted any correlation between ploidy and the number of nucleoli. He also stated that tumors have many more nucleoli than the normal organs of the hybrids. This conflicts with the findings of Hitier and Izard (1951). These workers failed to find an unusually high number of nucleoli in tumor tissue; they found one to three per nucleus, the same number as found in the meristems of their amphidiploids.

Whitaker (1934) considered that the introduction of paternal *N. langsdorffii* chromosomes into *N. glauca* cytoplasm caused the abnormal proliferations. This theory is difficult to reconcile with the fact that all sorts of combinations, including the reciprocal crosses, form tumors (Kehr and Smith 1954). The relation of nuclear abnormalities to tumor formation also remains unclear. Brieger and Forster (1942) point out that the derivation of tumors from mature cells excludes the possibility that abnormal mitoses cause directly the formation of tumors. Kehr and Smith (1954), on the other hand, conclude that tumorous development is at least associated with abnormal cell division and that the condition which ultimately brings about the development of tumorous growth is also the condition which results in abnormal mitoses in the root tips.

Genetic aspects

The first serious attempt to analyze the precise genetic basis of these tumors was made by KEHR and SMITH (1952, 1954). These authors bred a considerable number of diploid and polyploid combinations of *Nicotiana glauca* × *N. langsdorffii*. Genome combinations were obtained which ranged from an extreme of two *N. langsdorffii* genomes to one of *N. glauca* (LLG, a tumor-bearing combination also produced by HITIER and IZARD 1951) to the other extreme of one genome of *N. langsdorffii* to three genomes of *N. glauca* (LGGG). Tumors were conspicuous on all plants in all these combinations. From these data it was concluded that the tumor-forming nature of the hybrid *N. glauca* × *N. langsdorffii* remains relatively unchanged regardless of the ratio of *N. glauca* and *N. langsdorffii* genomes as long as at least one complete genome of each species is present in the hybrid. No exception to this rule was found even in multiple genome combinations involving up to four species. It was possible for KEHR and SMITH to dilute out the *N. glauca* genome and to obtain a majority of the twelve *glauca* chromosomes as extra single chromosomes or in combination with other extra *glauca* chromosomes on an essentially diploid *N. langsdorffii* background (LL + some G. chromosomes) by using the following genetic technique. *N. langsdorffii* (LL) and the sesquidiploid (LLG) possessing two *langsdorffii* genomes and one *glauca* genome were crossed. This was followed by the selection of offspring which showed the most pronounced *N. glauca* characteristics. Such offspring were selfed or backcrossed to *N. langsdorffii*. None of these plants formed tumors. In other words, spontaneous genetic tumors develop in the presence of all twelve *N. glauca* chromosomes (the complete genome) added to a diploid complement of *N. langsdorffii*, but no tumors develop when only one or a few *glauca* chromosomes are present in addition to the diploid *langsdorffii* genome.

The problem has recently been approached from a different direction by NÄF (1958). He has divided the parents of tumorous hybrids into two groups which he has arbitrarily designated as "plus" and "minus." The former comprises the Alatae and perhaps the Noctiflorae sections of the genus; the latter, all other species shown to be involved in tumor formation. If an intragroup cross is made either between two plus species or two minus species, the offspring will not develop tumors. On the other hand, crosses between a plus and a minus species produce tumorous offspring. There are very few exceptions to this scheme and NÄF explains these on the basis of relative plusness and relative minusness, similar to HARTMANN's concept of relative sexuality. That is, a weak minus species may not react with a weak plus species but may react with a very strong minus species. NÄF, on the basis of his theory, was able to predict and obtain new tumorous, as well as previously unreported non-tumorous, hybrid combinations. He concluded that the critical contributions of minus parents differ from those of plus parents. These contributions, while primarily of a genetical nature, should also be reflected in the parental metabolism. Accordingly, attempts are now being made to characterize these on a physiological level.

In vitro studies

White (1944) cultured fragments of *Nicotiana langsdorffii* × *N. glauca* tissue for 5 years, then grafted them into stems of *N. glauca*. In about one-third to one-half of the cases, the implants developed into tumors. This may be considered as a satisfactory criterion of tumor tissue. On the other hand, this tissue can be made to differentiate. Ordinarily, such tissue is grown on semi-solid media and does not differentiate. When, however, such tissue, initially isolated 2 years previously and subcultured, was immersed in a liquid medium, it differentiated into shoots and leaves (White 1939). That is, placing the tissue under about 8 mm. of liquid caused the tumorous tissue to differentiate mesophyll and palisade tissue, epidermis with stomata and hairs, and vascular tissue. It was presumed that oxygen gradients were instrumental in the differentiation process. After 144 passages the tissue isolated originally by White still differentiated in liquid culture (Skoog 1944). In 1953, after an estimated 300 passages, Kehr and Smith (1954) were unable to obtain any differentiation in a liquid medium. A more recently isolated comparable tissue did, at that time, differentiate under comparable cultural conditions. It is of interest to note that the tumorous tissues differentiate shoots and leaves only, never roots, except secondarily from shoots. This parallels the behavior of the tumor tissue *in vivo*.

Skoog (1944) has reported that the tendency to form buds and shoots can be completely suppressed by the addition of 0.2 parts per million of indoleacetic acid. This IAA effect can be reversed, however, by raising the level of certain nutrients such as KH_2PO_4, $Fe_2(SO_4)_3$, and sucrose. In liquid culture a low concentration of IAA not only suppresses differentiation but also stimulates growth somewhat, so that final fresh weights may be about fifty per cent in excess of that of controls.

Tumor tissue grows well in culture in the absence of growth hormones of the auxin type, as does tissue isolated from organized stems, *i. e.* stems of the hybrid which have not as yet proceeded to form tumors. Such cultured tissue when grafted into *N. glauca* stems produces tumors (Kehr and Smith 1954). Kehr and Smith suggest that the process by which this organized tissue converts to tumor tissue may be analogous to that of habituation, or, alternatively, that a dilution of some hormone, inhibitor, or other vital factor results in the formation of tumorous tissue.

Auxin assays performed by Skoog (1944) on White's original strain of *N. glauca* × *N. langsdorffii* tumor tissue indicated the following. No auxin was obtained by diffusion into agar blocks but ether extractions yielded considerable amounts. An ether-soluble, water-soluble substance was also found which decreased the activity of auxin as determined by the *Avena* test. The inhibitory substance increased as the cultures reached a point at which they nearly ceased to grow. More recently, HENDERSON, in collaboration with KEHR and SMITH (KEHR and SMITH 1954), reported that the leaf tissue of *N. glauca* × *langsdorffii* contains more free tryptophan, more free auxin, and is more efficient in converting tryptophan to auxin than either of the parents from which it was derived.

Discussion of genetic tumors

As in the other two neoplastic diseases discussed above, in the *Nicotiana* hybrid tumors the apparent primary agent which is responsible for the formation of tumors, in this case the genetic constitution of the cell, does not elicit tumor formation unless some other condition is also fulfilled. This second condition is satisfied when cells in mature tissue are stimulated to divide. The stimulus may be exerted by an external agent such as mechanical, chemical, or radiation damage, or it may be the consequence of normal, internal processes such as leaf abscission, or perhaps lateral root initiation. That is, all mature cells capable of division may be considered as incipient tumor cells and need only be stimulated to divide to unmask the neoplastic properties.

A second principle demonstrated by the genetic tumors is that of "autonomy" in respect to growth hormones. As in the case of crown gall and BLACK's virus tumor, tissue isolated from the hybrid tumors is capable of synthesizing from mineral salts and sucrose all of the growth factors required for its continued abnormal growth. What is particularly interesting in this case is the fact that cells obtained from non-tumorous portions of hybrid stems also become autonomous upon isolation and culture (KEHR and SMITH 1954). As indicated above, such auxin-independent tissue is truly neoplastic since it produces tumors upon implantation into healthy hosts.

It would seem almost superfluous to state that a number of theories have been proposed to account for the phenomenon of genetic tumors in *Nicotiana,* and that none of the theories can account satisfactorily for all aspects of the phenomenon. KEHR (1951) has critically reviewed the theories proposed by earlier workers.

In his earlier papers KOSTOFF (1930, 1933) made much of certain precipitation reactions which he had observed to occur when mixing juices of paternal and maternal species. He even implied that in human cancer the mixing of blood types bore a direct relation to the incidence of cancer. It may be possible that the genetic make-up of the hybrid cell leads to the precipitation of certain components at certain stages in development, but in the light of present-day knowledge, the more general precipitin reaction invoked by KOSTOFF is highly improbable. Grafts between *N. langsdorffii* and *N. glauca* fail to produce tumors.

In 1943 KOSTOFF proposed a theory of tumor induction based on abnormal mitoses. The fact that nuclear abnormalities probably represent a secondary phenomenon has already been discussed. KEHR (1951) provides additional evidence by his finding that in the tumor-bearing amphidiploid

N. glauca × *N. langsdorffii* normal meiotic divisions were found in more than 99 per cent of the pollen mother cells examined. As Kehr points out, it would be logical to expect abnormal meiotic divisions in the sex cells if abnormal mitotic divisions were a regular and frequent occurrence in the apical meristem.

Kehr (1951) proposes a theory based on abnormal phytohormone relationships. His hypothesis states that in certain genome or chromosome combinations of *Nicotiana* species, particularly those involving *N. langsdorffii,* the genetically-controlled phytohormone metabolism, which in the species themselves produces normal growth, is disturbed. This disturbance in the growth-regulatory balance stimulates certain tissues of the hybrids to divide, resulting in the formation of undifferentiated masses of plant material. This last hypothesis appears to be the most reasonable. However, some of the more specific aspects of the theory as put forth by Kehr (1951) in terms of five stated assumptions need to be modified: *1)* plants tend to produce tumors principally during a period of reduced terminal meristematic growth (page 59), but they can be made to do so at any time by wounding near a vascular bundle (Kunkel 1954). *2) Nicotiana langsdorffii is* more effective in tumor-forming hybrids than any other species of *Nicotiana* tested (page 58). This observation may reflect as much the range of species with which *N. langsdorffii* may be crossed as its potentiality to cause tumor formation upon hybridization. Furthermore, the *N. glauca* genome, both by itself and in combination with other genomes, produces more tumor-bearing hybrids than does *N. langsdorffii* (Brieger and Forster 1942; Kehr and Smith 1954). *3) N. langsdorffii* possesses a genetically-controlled ability to make hormones less effective (page 58). This may or may not be correct. Unfortunately, the basis for this statement involves a comparison between the ability of the corollas of *N. alata* and *N. langsdorffii* to respond by further cell elongation to the application of IAA. These two species do not form tumors when crossed. According to Näf both these parents belong to the plus group and therefore both make the same critical contribution to tumor formation in hybrids. In attempting to define the precise contribution made by each parent to tumor formation it is more pertinent to define the differences between plus (*e. g. langsdorffii,* *N. alata*) and minus (*e. g. N. glauca, N. suaveolens*) parents than differences between species of the same group. Kehr's other two assumptions are well established, *i. e., 1)* that hormone metabolism in plants is controlled by gene action, and *2)* that hormones may produce overgrowths. However, even the last statement needs to receive a certain amount of reconsideration and reinterpretation.

It is true that applications of hormones of the auxin type may produce disorganized self-limiting overgrowths, and that perhaps the genetic constitution of the hybrid results in an excess production of such hormones. It would be a mistake, however, to focus attention on auxin autonomy alone when looking for the physiological basis of the growth autonomy exhibited by the hybrid tumor tissue. Firstly, the auxin autonomy is not complete since the addition of 2×10^{-7} indoleacetic acid to the medium exerts a stimulatory effect (SKOOG 1944). Secondly, both the histology and the morphology of the tumors belie the possibility that the tissue is subjected simply to an excess of auxin.

Generally the hybrid tumors are more organized and differ histologically from crown gall in two important respects. In the growth of the hybrid tumors there is usually relatively little involvement of the cambium, and vascularization is slight. It is known that both of these phenomena may be induced by auxin (SNOW 1935; JACOBS 1952). In crown gall the cambium proliferates and lays down a considerable amount of xylem. In addition, one of the most striking symptoms of more developed crown gall is the bizarre knotting of tracheids found in the disorganized tumor tissue. In hybrid tumors vascularization is much lighter and there is usually no evidence of any markedly irregular arrangement of the fundamental vascular tissues. LEVINE (1937) has pointed out that the early death of certain of the tumors suggests that the vascularization is inadequate. Furthermore, auxin-induced overgrowths show a histological picture which resembles crown gall, not the hybrid tumors, although such studies were never made on the appropriate *Nicotiana* hybrids (*e. g.* KRAUS, BROWN and HAMNER 1936; BECK 1949).

One of the most characteristic aspects of the morphology of the hybrid stem tumors is their tendency to form small, aberrant, occasionally normal-appearing leaves and shoots. The tendency on the part of the hybrid tumor tissue to produce buds and shoots when grown under a layer of liquid can be completely suppressed by the addition of 2×10^{-7} IAA to the medium (SKOOG 1944).

All attempts to organize exhibited by the hybrid tumors involve the production of shoots and leaves only, roots are never produced. This is true *in vitro* as well as *in vivo*, except where roots are secondarily formed from tissues of the intact plant. As is well known, roots may be initiated by the application of high concentrations of auxin. In crown gall, spontaneous root production is a characteristic of many plants and is occasionally encountered in tobacco. In other words, those morphological symptoms which reflect a state of hyperauxiny such as root formation and the suppression of shoot formation and which are characteristically found to be present in crown gall, are absent in the hybrid tumors.

It can be seen, then, that not only does the growth of the hybrid tumor tissue respond to added auxin, but both the histology and morphology of the tumors indicate that the tissue is not in a state of auxin-excess comparable to that encountered in crown gall. That the term auxin-excess does not involve an absolute quantity of auxin but rather a ratio of auxin to some other factor is indicated by the reports of SKOOG and co-workers. The fact that the IAA suppression of liquid culture differentiation of the hybrid tumor tissue can be reversed by raising the level of certain nutrients has already been mentioned (SKOOG 1944). Subsequent studies on tobacco (SKOOG and TSUI 1951; MILLER and SKOOG 1953) demonstrated that bud formation on isolated stem segments depends upon an adenine : indoleacetic acid ratio. That is, the addition of adenine fosters bud development, a process which may be inhibited by the addition of IAA. The IAA inhibition, in turn, can be overcome by the addition of larger amounts of adenine. More recently Skoog and Miller (1957) have reported that 6-furfurylaminopurine (kinetin) is much more effective than adenine in this respect. It is interesting to note that certain tissues such as pith of *Nicotiana tabacum* require kinetin in addition to IAA for growth (Miller et al. 1955 a) and that in crown gall induction of tobacco pith both growth-substance synthesizing systems become activated (Braun 1956 a). Furthermore, stem tissue derived from Näf's plus parents needs other factors in addition to auxin for continued growth *in vitro* (Näf 1958). These considerations, then, suggest that the growth of the genetic tumors depends not solely on an excess of indoleacetic acid but that other growth substances play as important or perhaps even more important a role in achieving growth autonomy.

General Discussion

A century of experience has made it abundantly clear that autonomy and anaplasia are the distinguishing biological characteristics of the cancer cell. True tumor cells of plants are altered, randomly proliferating cells that reproduce true to type and against the growth of which there is no control mechanism in the host. These abnormal plant cell types, like animal cancer cells, exhibit both autonomy and anaplasia.

Plant tumors may assume a variety of growth patterns. These range from slowly growing. benign to rapidly growing malignant, and from completely unorganized to highly organized growths. All may be produced under controlled experimental conditions and all have their counterparts in animal pathology.

Problems concerned with the nature of autonomy as well as those concerned with the morphological, histological, and cytological peculiarities that characterize the neoplastic state are fundamental and constitute the ultimate basis of the tumor problem. These are problems of growth and their solution will, in all likelihood, show common features in animals and plants, for at the cellular level members of the two kingdoms have much in common.

The most important single concept that arises from the plant work is that dealing with the nature of autonomy. Oncologists have long asserted that the unregulated growth of tumor cells reflects an independence of those cells from the morphogenetic restraints that govern the growth of normal cells within an organism. What precisely constitutes the morphogenetic restraints placed upon a normal cell and what is entailed in overcoming restraints under conditions of neoplasia are not understood. Theoretically, autonomy of cancer cells requires within them something newly activated and distinctive, something that urges those cells to continued abnormal proliferation. Many hypotheses of the most diverse types have been advanced over the years to account for the continuing cause or causes of malignancy. Of particular interest to this discussion was the suggestion made by LEO LOEB (1937, 1945) somewhat more than a decade ago. LOEB, one of the pioneers of experimental oncology, hypothesized that, in the case of animal cancer cells, a growth substance is produced or increased in quantity and that the production of this substance is renewed auto-catalytically. This suggestion was based in part on the observation that the parenchymal cells of tumors may stimulate their stroma to neoplasia, indicating that tumor cells produce excessive amounts of a growth-promoting substance. There is now an abundance of evidence to indicate that this is precisely what happens when normal plant cells are converted to tumor cells. Plant cells acquire as a result of their transformation a capacity to synthesize greater than regulatory amounts of growth substances of the type concerned with growth accompanied by cell division. The continued production in greater than regulatory amounts of such substances can and most probably does account for the continued abnormal proliferation of plant tumor cells. The degree to which a cell is altered appears, moreover, to be a reflection of the extent to which the growth-substance synthesizing systems are activated during the transformation process.

The student of plant growth has available to him chemically pure substances with which he can control under precisely defined experimental conditions such fundamental growth processes as cell enlargement and cell division. By varying the concentrations of these two substances in a chemically defined culture medium, it has been possible to reproduce not only all the morphological growth patterns (disorganized tumors and teratomata formation) but also the histological (hypertrophy and hyperplasia leading to disorganization and loss of function) and cytological (multinucleate giant cells, aberrant nuclear behavior, etc.) abnormalities that are commonly observed in certain tumor tissues. It is thus possible to account at a physiological level not only for autonomy but for anaplasia as well. These artificially stimulated normal cells are self-limiting and, when the growth substances are removed, their growth promptly stops. The fact that such stimulated normal cells commonly show histological and cytological characteristics of true tumor cells but are themselves self-limiting indicates that the observed cellular abnormalities are the result rather than the cause of the tumorous state.

Although the tumor cells themselves elaborate the growth-promoting substances which urge these cells on to continued proliferation, other sources of the growth hormones may also be of etiological significance. This is particularly true when dealing with cells possessing relatively low grades of neoplastic change. Growth hormones supplied to such low grade tumors by the hormone-producing centers of the plant, or applied artificially, may influence both the morphological growth patterns and the rates of growth of the resulting tumors. When, for example, the tumor-inducing principle elaborated by a moderately virulent strain of the crown gall bacterium transforms pluripotent cells at the cut stem tip of a tobacco plant, complex tumors or teratomata arise. When, on the other hand, this same tumor-inducing principle alters similar cells in an internode of a plant containing a functional apical bud, and hence a hormone-producing center, typical unorganized crown gall tumors develop. In this instance the effect of the functional apical bud can be wholly replaced by synthetic growth hormones. Not only the tendency to organize but also the rate of growth may be hormonally influenced. The tumor-inducing principle associated with attenuated strains of the inciting bacteria initiates very slowly growing tumors in plants such as the tomato. The application of synthetic hormones of the auxin type to such normally slowly growing tumors results in the formation of large overgrowths, the rapid growth of which continues only as long as the tumor tissue is supplied with the growth-promoting substance. In this instance, as in that reported above, the effect of the growth substances on the cell is only temporary and, although growth is enhanced, the increased growth rate does not become an intrinsic property of the cell itself. It therefore follows that the greater the degree of the primary change and hence the greater the growth-substance output by the tumor cell itself, the less effective are externally supplied hormones in stimulating growth. The growth of fully altered rapidly growing crown gall cells is, in fact, inhibited when such cells are treated with hormones of the auxin type. The stimulatory as well as the inhibitory effects of hormones on tumor growth in animals have been commonly noted.

The ultimate mechanism by which growth-substance synthesizing systems become activated in plant tumor cells is unknown. The production of these growth substances by the cell is presumably enzymatic in nature. Since enzymatic reactions are commonly considered to be gene-controlled processes, it would appear that the normal gene complement is somehow modified in the plant tumor cell. This could conceivably be accomplished by somatic mutation at the genic level. The concept of somatic mutation as a possible explanation for the origin of a tumor cell was first suggested by Boveri (1929) shortly after the turn of the century. This theory has seemed particularly attractive to some investigators, not only because of the frequent association of cytological abnormalities with the cancerous state, but also because it appears to offer a most logical explanation for the multiplicity of diverse agencies that have been reported to be concerned in tumor genesis. Several forms of radiant energy as well as various chemical substances have been found to possess highly effective mutagenic as well as carcinogenic properties.

There are several reasons for qualifying the somatic mutation hypothesis, especially as it concerns the plant tumor work. The evidence indicates, first of all, that the cytological anomalies found associated with the tumorous state are secondary effects of an upset growth-hormone metabolism in the tumor cells. Secondly, some investigators have interpreted the diverse agencies found to be concerned with the genesis of tumors as involving the unmasking of latent tumorogenic viruses. This interpretation is strengthened by the fact that specific viruses can regularly initiate tumors in animals and plants, and in some instances such viruses are masked. Since this is true, one may question whether the tumor cell is necessarily an irreversibly altered cell. It would not, for example, be difficult to conceive in the case of BLACK's virus tumor that the elimination or inactivation of the inciting virus would return the cells to normalcy. Although this has not yet been done in this particular instance, other closely related viruses that produce abnormal growth patterns in plants have been eliminated from their hosts by thermal treatment with a resulting complete recovery of such plants (KUNKEL 1936, 1941). Plant viruses of this type apparently do not, therefore, induce permanent modifications in plant cells. The continued abnormal behavior of the cell appears to be dependent upon the continued presence of the virus.

If in the plant virus tumors recovery of the tumor cells has not yet been demonstrated, in the crown gall disease of plants the controlled recovery of tumor cells is believed to have been accomplished experimentally. Crown gall tumor cells of the commonly found type possess a high degree of neoplastic change and have generally been regarded as being permanently altered cells. They were therefore unsuited for studies on recovery. It was found, however, that when the tumor-inducing principle associated with a moderately virulent strain of the inciting bacterium in crown gall transformed pluripotent cells that possessed at the time of their alteration highly developed regenerative capacities, complex tumors or teratomata resulted. These were characterized both in the host and in culture by a capacity to organize highly abnormal tissues and organs. By forcing such abnormal tumor buds into very rapid growth by a series of graftings to healthy plants, a gradual but ultimately complete recovery was achieved. These results make somatic mutation at the genic level as a possible explanation of the nature of the cellular alteration in crown gall appear highly unlikely. They suggest rather that the factor responsible for the continued abnormal proliferation of the crown gall tumor cell is an autonomous or a partially autonomous entity that is subject to the effects of dilution in cells that are forced to divide with great rapidity. In considering these findings it should be recalled that there exists a basic difference in the regenerative capacity of cells of higher animals and plants. The character of somatic cells of higher animals is often specifically determined and strongly fixed very early in their development, while somatic cells of certain higher plants may and frequently do retain a high degree of pluripotency. A single plant cell, or a small group of cells, under the proper stimulus is capable of regenerating an entire plant. Plant cells behave in this respect as do cells of certain lower animals.

The findings reported above suggest, then, that crown gall tumor cells may recover if they are forced to divide with unusual rapidity at the stem apex. These results are reminiscent of those encountered in microbial genetics. Studies such as those presented by EPHRUSSI (1951) and SPIEGEL-MAN (1954) have made it abundantly clear that certain self-duplicating cytoplasmic factors, as well as the nuclear genes, may serve as determinants of heritable differences in a cell. These cytoplasmic entities appear to be concerned in enzyme production. Therefore, mechanisms quite different from that of somatic mutation at the genic level can be postulated to explain the continuity of cancerous properties from one cell generation to the next.

Although somatic mutation at the genic level does not appear adequate to explain the continued abnormal growth of the plant tumor cell in crown gall and in Black's virus tumor, there seems to be no question about the fact that the genetic constitution of a cell may play both a primary and a secondary role in determining whether tumors will be elicited. The primary contribution to the tumorous state by the genome of a cell is perhaps best exemplified in the case of Kostoff's hybrid tumors. In the absence of irritation, these hybrid plants are for the most part perfectly organized both morphologically and histologically during their period of active growth. When, however, such plants reach maturity, a profusion of tumors invariably appears in all parts of the plant. The tumors may be induced earlier if the plants are wounded close to a vein. These hybrids appear, then, to be composed entirely of potential tumor cells that are maintained in a state of perfect organization during the period of their active growth. The cells of such plants behave as do normal cells until they are stimulated to divide as a result of either natural processes or artificially-induced irritation. Once the cells of this plant are activated, they no longer respond to the morphogenetic restraints that return a normal cell to quiescence. The genetic constitution of the cell, in this instance, appears to be critical. Only such a non-specific stimulus as irritation is required to transform the potential tumor cells, of which the hybrid plant is composed, into actively proliferating autonomous plant cell types.

The case of the hybrid tumors is not, of course, an example of somatic mutation. As indicated, the basis for the tumor formation is the genetic constitution of the plant (*i. e.* presumably of all of its cells), although its expression is limited to cells reacting to appropriate stimuli. A phenomenon which may turn out to be a case of somatic mutation, modified in the sense of the previous discussion, is that of habituation. The manner in which habituated tissue arises is very suggestive. Cells, which suddenly exhibit growth hormone-independence, appear to arise as a sector in the clump of cells which comprises a piece of cultured tissue. No external agent has been associated with this event, and it appears to occur spontaneously and with a very low frequency. Unfortunately, so little is understood of this phenomenon at this time that invoking the somatic mutation hypothesis as an explanation must, perforce, be very tentative. Gautheret, who is perhaps most familiar with this phenomenon, has, in fact, suggested that it represents some sort of an enzymatic adaptation rather than a somatic mutation at the genic level.

If, as indicated above, the genetic constitution of a cell may play a primary role in carcinogenesis, it may also play an important, if secondary, role in the formation of tumors whose primary cause is non-genetic. This is well illustrated in BLACK's virus tumor where heredity influences the disease in at least two ways. The genetic constitution of the host may determine susceptibility to the tumor-inducing virus *per se* or it may determine the response of cells to the presence of the virus. Thus the virus may 1) fail to multiply in a plant, 2) multiply but fail to elicit tumors, 3) multiply and initiate the formation of tumors. In the third category, within the same plant species various clones may show striking differences with respect to frequency, distribution, size and shape of the resulting tumors. It is highly significant that one clone which responds readily to virus infection with tumor formation also occasionally gives rise to tumors spontaneously in the non-infected state. This situation is comparable to that observed with strain C_3 Hb of mice that have lost the mammary carcinoma virus yet show an inherent tendency to develop mammary carcinoma.

Since certain viruses appear to be the only agencies yet characterized that are capable of accounting for the continued abnormal growth of tumor cells, it has been argued by some investigators that all tumors are caused by viruses, some of which are masked and hence not easily demonstrable. Just as the fact that viruses can elicit tumor formation in plants and animals makes the mutation hypothesis untenable as being *the* sole cause of the cancerous change, so the fact that the KOSTOFF hybrid tumors exist makes unlikely the postulate that all tumors are caused by viruses. If, nevertheless, it could be established experimentally that all neoplastic diseases did indeed involve a virus, it still would not explain enough. The question remains *how* does a virus induce the neoplastic state in the host cell. Viruses cause cells to respond to infection in many different ways. Few elicit tumorous growths. Even the carcinogenic viruses may infect cells and not exert a stimulatory influence on such cells. What, then, causes these tumor viruses to induce cells to proliferate independently of the morphogenetic restraints that govern the growth of all normal cells within an organism? There is a clue in the case of BLACK's virus tumor. The virus-infected tumor tissue shows a capacity to synthesize growth substances of the type concerned with growth and cell division in amounts adequate to achieve autonomy. It thus appears that the presence of the virus confers upon the cell the ability to produce such growth substances in greater than regulatory amounts. But as soon as this is postulated it must be conceded that agencies other than viruses might also produce a similar end result. The evidence in the field of plant oncology implies that the physiological autonomy which underlies the tumorous state may be achieved by at least three and perhaps four different and quite distinct means.

The role of irritation in carcinogenesis has remained obscure. At one time irritation itself was thought to be the cause of cancer. This hypothesis has now been largely discarded. On the basis extensive clinical and experimental observations it is nevertheless generally recognized that

6*

irritation does constitute an important contributory factor. In crown gall of plants the contribution of wounding to tumor-induction is clear in at least one respect. The wound, in this instance, is essential to make host cells vulnerable to the carcinogenic action of the tumor-inducing principle. This is based on the observation that the presence or absence of the tumor-inducing principle is irrelevant to tumor initiation except at a certain specific time following wounding. It is only after wound healing has been allowed to take place for about 34 hours that it is possible to transform the host cells to tumor cells. An optimum vulnerability appears to take place 60 hours post-wounding and in most plants disappears after about 5 days.

In the case of virus tumors and genetic tumors, the role of wounding is more obscure and less specific than in crown gall, since other stimuli such as hormone application, lateral root formation, leaf abscission, etc.. may be equally effective. It appears unlikely in these instances that the cells need to be made vulnerable to the primary carcinogenic agent. The virus infection may be systemic long before tumors are initiated by wounding. In the case of the hybrid tumors, the genetic constitution peculiar to that type of cell is an intrinsic property of the cell from the time of its conception. What appears to be more likely, therefore, is that the primary neoplastic agent does not induce quiescent cells to divide but rather that the induced tumorous state prevents proliferating cells from returning to quiescence. That is, tumor cells in these instances do not develop into neoplastic growths unless mature cells are first stimulated to divide by non-carcinogenic processes such as by wounding, application of hormones, etc.

In summation, the following picture emerges from the plant work. Several quite distinct agencies can bring about the tumorous state. The basic biological aspects of this state are characterized by a growth autonomy and a tendency to anaplasia. Both these phenomena can be explained on the basis of the ability of the tumor cell to synthesize growth hormones in greater than regulatory amounts. Whether this growth-hormone autonomy is brought about by an enhancement of the hormone-synthesizing apparatus itself, or by a change resulting in the decreased efficiency or even total elimination of a system regulating the hormone level, is not known.

Acknowledgements

This review was written while the junior author held a Damon Runyon Memorial Cancer Research Fellowship.

The authors are pleased to acknowledge their indebtedness to Drs. L. O. Kunkel, U. Näf, R. C. King, K. Maramorosch, and V. C. Littau for their critical reading of various parts of the manuscript, to Mr. J. A. Carlile for the illustrations, and to Miss E. J. Ross for invaluable help in the preparation of the manuscript.

Bibliography

APPLER, H.. 1951: Über die Tumorbildung durch *Pseudomonas tumefaciens* und durch Heteroauxin an *Helianthus annuus.* Biol. Zbl. 70, 452—469.

AULER, H., 1924: Zur Histogenese der Tumefaciensgeschwülste an der Sonnenblume. Z. Krebsforsch. 21, 354—360.

BANFIELD, W. M., 1935: Studies in cellular pathology. I. Effects of cane gall bacteria upon gall tissue cells of the black raspberry. Bot. Gaz. 97, 193—239.

BEARDSLEY, R. E., 1955: Phage production by crown-gall bacteria and the formation of plant tumors. Amer. Naturalist 89, 175—176.

BECK, E. G., 1949: The effect of wounding and heteroauxin on the development of crown gall on *Impatiens balsamina.* Amer. J. Bot. 36, 793—794.

BENDER, E., und W. BRUCKER, 1956: Studien zur zellfreien Tumorübertragung an Pflanzen I. Z. Bot. 44, 531—542.

BERGEY's Manual of Determinative Bacteriology, 6th ed., 1948, Williams & Wilkins Co., Baltimore, Md.

BERRIDGE, E. M., 1930: Studies in bacteriosis. XVII. Acidic relations between the crown-gall organism and its host. Ann. Appl. Biol. 17, 280—283.

BITANCOURT. A. A., 1949: Mecanismo Genético da Tumorisação nos Vegetais. Segunda Semana de Genética, Piracicaba, São Paulo, Feb. 8—12, 1949.

— 1954: La nature des auxines des tumeurs végétales. L'Année Biologique, 3e sér. 30, 361—370.

BLACK, L. M., 1944: Some viruses transmitted by agallian leafhoppers. Proc. amer. Philos. Soc. 88, 132—144.

— 1945: A virus tumor disease of plants. Amer. J. Bot. 32, 408—415.

— 1949: Virus tumors. Survey of Biological Progress 1, 155—231.

— 1951: Hereditary variation in the reaction of sweet clover to the wound-tumor virus. Amer. J. Bot. 38, 256—267.

— 1952: Plant virus tumors. Ann. N. Y. Acad. Sci. 54, 1067—1075.

— 1953: Occasional transmission of some plant viruses through the eggs of their insect vectors. Phytopathology 43, 9—10.

— 1954: Plant tumor diseases and the relation of some of them to viruses. Proc. 2nd Nat. Cancer Conf., Ohio, Mar. 3—5, 1952, 2, 1349—1355.

— 1955: Concepts and problems concerning purification of labile insect-transmitted plant viruses. Phytopathology 45, 208—216.

— and M. K. BRAKKE, 1952: Multiplication of wound-tumor virus in an insect vector. Phytopathology 42, 269—273.

— and C. L. LEE, 1957: Interaction of growth-regulating chemicals and tumefacient virus on plant cells. Virology 3, 146—159.

— K. MARAMOROSCH, and M. K. BRAKKE, 1950: Filtration and sedimentation of wound-tumor virus. Phytopathology 40, 2.

BLOCH. R., 1941: Wound healing in higher plants. Bot. Rev. 7, 110—146.

— 1952: Wound healing in higher plants. II. Bot. Rev. 18, 655—679.

BONNER, W. D., Jr., 1957: Soluble oxidases and their functions. Ann. Rev. Plant Physiol. 8, 427—452.

BOVERI, T., 1929: The origin of malignant tumors. English translation by Marcella Boveri. Williams & Wilkins Co.. Baltimore, Md.

BRAKKE, M. K., K. MARAMOROSCH. and L. M. BLACK, 1953: Properties of the wound-tumor virus. Phytopathology 43, 387—390.

— and L. G. NICKELL, 1951: Secretion of α-amylase by *Rumex* virus tumors *in vitro.* Properties and assay. Arch. Biochem. Biophys. 32, 28—41.

— — 1955: Secretion of an enzyme from intact cells of a higher plant tumor. L'Année Biologique. 3e sér. 31, 215—226.

— A. E. VATTER and L. M. BLACK. 1954: Size and shape of wound-tumor virus. Brookhaven Symp. Biol., No. 6, Abnormal and Pathological Plant Growth, 137—156.

BRAUN, A. C., 1941: Development of secondary tumors and tumor strands in the crown gall of sunflowers. Phytopathology 31, 135—149.

— 1943: Studies on tumor inception in the crown-gall disease. Amer. J. Bot. 30, 674—677.

— 1947 a: Thermal studies on the factors responsible for tumor initiation in crown gall. Amer. J. Bot. 34, 234—240.

— 1947 b: Recent advances in the physiology of tumor formation in the crown-gall disease of plants. Growth 11, 325—337.

Braun, A. C., 1948: Studies on the origin and development of plant teratomas incited by the crown-gall bacterium. Amer. J. Bot. 35, 511—519.
— 1950: Thermal inactivation studies on the tumor-inducing principle in crown gall. Phytopathology 40, 3.
— 1951 a: Recovery of crown-gall tumor cells. Cancer Research 11, 839—844.
— 1951 b: Cellular autonomy in crown gall. Phytopathology 41, 963—966.
— 1952: Conditioning of the host cell as a factor in the transformation process in crown gall. Growth 16, 65—74.
— 1953: Bacterial and host factors concerned in determining tumor morphology in crown gall. Bot. Gaz. 114, 363—371.
— 1954 a: The physiology of plant tumors. Ann. Rev. Plant Physiol. 5, 133—162.
— 1954 b: Studies on the origin of the crown-gall tumor cell. Brookhaven Symp. Biol., No. 6, Abnormal and Pathological Plant Growth, 115—127.
— 1956 a: The activation of two growth-substance systems accompanying the conversion of normal to tumor cells in crown gall. Cancer Research 16, 53—56.
— 1956 b: A study on the nature of autonomy in neoplastic plant cells. Amer. Naturalist 90, 227—236.
— 1957 a: Tissue culture as a tool for studying the development of autonomy in neoplastic plant cells. Lect., Decennial Rev. Conf. on Tissue Culture, 1956. J. Nat. Cancer Inst. 19, 753—769.
— 1957 b: A physiological study on the nature of autonomous growth in neoplastic plant cells. Soc. exper. Biol., Symp. XI, The Biological Action of Growth Substances, 132—142.
— and T. Laskaris, 1942: Tumor formation by attenuated crown-gall bacteria in the presence of growth-promoting substances. Proc. Nat. Acad. Sci. (U. S.) 28, 468—477.
— and R. J. Mandle, 1948: Studies on the inactivation of the tumor-inducing principle in crown gall. Growth 12, 255—269.
— and G. Morel, 1950: A comparison of normal, habituated, and crown-gall tumor tissue implants in the European grape. Amer. J. Bot. 37, 499—501.
— and U. Näf, 1954: A non-auxinic growth-promoting factor present in crown gall tumor tissue. Proc. Soc. exper. Biol. a. Med. (Am.) 86, 212—214.
— and P. R. White, 1943: Bacteriological sterility of tissues derived from secondary crown-gall tumors. Phytopathology 33, 85—100.
Brieger, F. G., and R. Forster, 1942: Tumores em certos hibridos do genero Nicotiana. Bragantia 2, 259—274.
Brookhaven Symposia in Biology, No. 6, Abnormal and Pathological Plant Growth. Report of Symposium held August 3 to 5, 1953. Brookhaven National Laboratory, Upton, N. Y. 303 pp. 1954.
Brown, N. A., and F. E. Gardner, 1936: Galls produced by plant hormones, including a hormone extracted from Bacterium tumefaciens. Phytopathology 26, 708—713.
— and F. Weiss, 1937: Crown gall of the fasciated type on Asparagus sprengeri. U. S. Dept. Agric., Plant Dis. Reptr. 21, 31—32.
Burkholder, P. R., and L. G. Nickell, 1949: Atypical growth of plants. I. Cultivation of virus tumors of Rumex on nutrient agar. Bot. Gaz. 110, 426—437.
Butler, E. J., 1930: Some aspects of the morbid anatomy of plants. Ann. Appl. Biol. 17, 175—212.
Buvat, R., 1943: Phénomènes de dédifférenciation dans les tumeurs corticales produites chez la tomate. C. r. Acad. Sci. Par. 216, 127—129.
— 1945: Recherches sur la dédifférenciation des cellules végétales. II. Cultures de tissus et tumeurs. Ann. Sci. Nat. Bot., sér. XI, 6, 1—119.
Camus, G., et R. J. Gautheret, 1948: Sur la transmission par greffage des propriétés tumorales des tissus de crown-gall. C. r. Soc. Biol. 142, 15—16.
— S. G. Wildman, and J Bonner, 1951: Comparative study of the soluble proteins of crown gall and normal tissues of sunflower cultivated in vitro. Amer. Inst. Biol. Sci., Bull. 1, 34. (Title only; no text given.)
Chemin, E., 1937: Role des bactéries dans la formation des galles chez les floridées. Ann. Sci. Nat. Bot., Sér. X, 19, 61—72.
Cook, M. T., 1923: Early stages of crown gall. Phytopathology 13, 476—482.
Czosnowski, J., 1952 a: Physiological features of three types of tissue of Vitis vinifera: healthy, crown-gall, and chemical tumor, grown in vitro. Poznań. Towarz. Przyjaciół Nauk, Wydział Mat.-Przyrod., Prace Komisji Biol. 13 (no. 4). 189—208. (Transl. title.)

CzosNowski, J., 1952 b: Vitamin metabolism of plant tissues cultured *in vitro* on the basis of their activity towards growth substances of the auxin type. Poznań. Towarz. Przyjaciól Nauk, Wydzial Mat.-Przyrod., Prace Komisji Biol. 13 (no. 5), 209—246. (Transl. title.)

D'Amato, F., 1952: Polyploidy in the differentiation and function of plant tissues and cells. A critical examination of the literature. Caryologia 4, 311—358.

Dame, F., 1938: *Pseudomonas tumefaciens* (Sm. et Towns.) Stev., der Erreger des Wurzelkropfes, in seiner Beziehung zur Wirtspflanze. Zbl. Bakter. usw., Abt. II, 98, 385—429.

De Ropp, R. S., 1946: *In vitro* grafts. Nature 157, 628—629.

— 1947 a: The response of normal plant tissues and of crown-gall tumor tissues to synthetic growth hormones. Amer. J. Bot. 34, 53—62.

— 1947 b: The growth-promoting and tumefacient factors of bacteria-free crown-gall tumor tissue. Amer. J. Bot. 34, 248—261.

— 1947 c: The isolation and behavior of bacteria-free crown-gall tissue from primary galls of *Helianthus annuus*. Phytopathology 37, 201—206.

— 1948 a: The growth-promoting action of bacteria-free crown-gall tumor tissue. Bull. Torrey Bot. Club. 75, 45—50.

— 1948 b: The interaction of normal and crown-gall tumor tissue in in vitro grafts. Amer. J. Bot. 35, 372—377.

— 1948 c: The movement of crown-gall bacteria in isolated stem fragments of sunflower. Phytopathology 38, 993—998.

Elliott, C., 1951: Manual of bacterial plant pathogens. 2nd, entirely revised edition, Chronica Botanica Co., Waltham, Mass. 186 pp.

Ephrussi, B., 1951: Remarks on cell heredity. p. 241—262 in: Genetics in the 20th Century. Edited by L. C. Dunn. Macmillan Co., New York.

Esau, K., 1953: Plant Anatomy. John Wiley and Sons, Inc., New York. 735 pp.

Fogelberg, S. O., B. E. Struckmeyer and R. H. Roberts, 1957: Morphological variations of mitochondria in the presence of plant tumors. Amer. J. Bot. 44, 454—459.

Galston, A. W., J. Bonner, and R. S. Baker, 1953: Flavoprotein and peroxidase as components of the indoleacetic acid oxidase system of peas. Arch. Biochem. Biophysics 42, 456—470.

— and L. Y. Dalberg, 1954: The adaptive formation and physiological significance of indoleacetic acid oxidase. Amer. J. Bot. 41, 373—380.

Garrigues, R., 1943: Recherches cytologiques sur les tumeurs à *Phytomonas tumefaciens*. C. r. Acad. Sci. Par. 217, 235—237.

Gautheret, R. J., 1946: Comparaison entre l'action de l'acide indole-acétique et celle du *Phytomonas tumefaciens* sur la croissance des tissus végétaux. C. r. Soc. Biol. 140, 169—171.

— 1947 a: Existe-t-il un cancer des plantes? Université de Paris. Conference au Palais de la Découverté, Paris, France.

— 1947 b: Comparaison entre la structure des cultures de tissus normaux et des cultures de tissus de crown-gall de Topinambour. C. r. Soc. Biol. 141, 598—601.

— 1952: Cancer végétal et culture des tissus. Rev. Path. comp. et Hyg. gén. 52, 100—120.

— 1955: The nutrition of plant tissue cultures. Ann. Rev. Plant Physiol. 6, 433—484.

Genevès, L., 1946: Recherches sur la formation de deux catégories cellulaires dans les tumeurs corticales de la tomate. Rev. Gén. Bot. 53, 381—411.

Goodwin, R. H., and F. Kavanagh, 1948: Fluorescing substances in roots. Bull. Torrey Bot. Club 75, 1—17.

— — 1949: The isolation of scopoletin, a blue-fluorescing compound from oat roots. Bull. Torrey Bot. Club 76, 255—265.

— — 1952: Fluorescence of coumarin derivatives as a function of pH. II. Arch. Biochem. Biophysics 36, 442—455.

— and B. M. Pollock, 1954: Studies on roots. I. Properties and distribution of fluorescent constituents in *Avena* roots. Amer. J. Bot. 41, 516—520.

Grasselli, G., and G. Cova, 1956: Induzione di tumori in piante diploidi e poliploidi. L'Ateneo Parmense 27, 3—7.

Hamdi, H., 1930: Über die Histogenese, Bau und Natur des sogenannten Pflanzenkrebses und dessen Metastasen. Z. Krebsforsch. 30, 547—552.

Hamner, K. C., and E. J. Kraus, 1937: Histological reactions of bean plants to growth promoting substances. Bot. Gaz. 98, 735—807.

Heald, D., 1933: Manual of Plant Diseases. 2nd edition. McGraw-Hill Book Co., Inc., New York.

Henderson, J. H. M., 1954: The changing nutritional pattern from normal to habituated sunflower callus tissue *in vitro*. L'Année Biologique, 3ᵉ sér. 30, 329—348.

— and J. Bonner, 1952: Auxin metabolism in normal and crown gall tissue of sunflower. Amer. J. Bot. 39, 444—451.

Hendrickson, A. A., I. L. Baldwin, and A. J. Riker, 1934: Studies on certain physiological characters of *Phytomonas tumefaciens, Phytomonas rhizogenes* and *Bacillus radiobacter*. Part II. J. Bacter. (Am.) 28. 597—618.

Heston, W. E., M. K. Deringer, T. B. Dunn, and W. D. Levillain, 1950: Factors in the development of spontaneous mammary gland tumors in agent-free strain C₃Hb mice. J. Nat. Cancer Inst. 10. 1139—1155.

Hildebrand, E. M., 1942: A micrurgical study of crown gall infection in tomato. J. Agric. Res. 65, 45—59.

Hildebrandt, A. C., and A. J. Riker, 1947: Influence of some growth-regulating substances on sunflower and tobacco tissue *in vitro*. Amer. J. Bot. 34, 421—427.

— — 1949: The influence of various carbon compounds on the growth of marigold, Paris-daisy, periwinkle, sunflower and tobacco tissue *in vitro*. Amer. J. Bot. 36, 74—85.

Hitier, H., and C. Izard, 1951: Contribution à l'étude des tumeurs observées sur certains hybrides de *Nicotiana*. C. r. Acad. Sci. Par. 232, 877—879.

Izard, C., 1952: Étude des tumeurs spontanées des certains hybrides interspécifiques de *Nicotiana*. Thesis, University of Toulouse, France.

Jablonski, J. R., and F. Skoog, 1954: Cell enlargement and cell division in excised tobacco pith tissue. Physiologia Plantarum 7, 16—24.

Jacobs, W. P., 1952: The role of auxin in differentiation of xylem around a wound. Amer. J. Bot. 39, 301—309.

Jakowska, S., 1949: Effects of *Bacterium tumefaciens* on *Allium cepa*. Phytopathology 39, 683—705.

Jensen, C. O., 1910: Von echten Geschwülsten bei Pflanzen. Deuxième Conférence Internat. pour l'Étude du Cancer. Rapport. Paris. p. 243—254, in Den Kgl. Veterinaer- og Landbohøjskoles Serumlab. VII.

— 1918: Undersøgelser vedrørende nogle svulstlignende dannelser hos planter. Den Kgl. Veterinaer- og Landbohøjskoles Aarsskrift 1918, 91—143.

Jones, S. G., 1947: An anatomical study of crown gall tumors on the Himalaya giant blackberry (*Rubus procerus*). Phytopathology 37, 613—624.

Kandler, O., 1952: Über eine physiologische Umstimmung von Sonnenblumenstengelgewebe durch Dauereinwirkung von β-Indolylessigsäure. Planta 40, 346—349.

Kehr, A. E.. 1951: Genetic tumors in *Nicotiana*. Amer. Naturalist 85, 51—64.

— and H. H. Smith, 1952: Multiple genome relationships in *Nicotiana*. Cornell Univ. Agric. exper. Sta. Mem. 311, 1—19.

— — 1954: Genetic tumors in *Nicotiana* hybrids. Brookhaven Symp. Biol., No. 6, Abnormal and Pathological Plant Growth, 55—78.

Kelly, S. M., and L. M. Black, 1949: The origin, development and cell structure of a virus tumor in plants. Amer. J. Bot. 36, 65—73.

Klein, R. M., 1952: Nitrogen and phosphorus fractions, respiration, and structure of normal and crown gall tissues of tomato. Plant Physiol. 27, 335—354.

— 1953: The probable chemical nature of crown-gall tumor-inducing principle. Amer. J. Bot. 40, 597—599.

— 1954: Mechanisms of crown-gall induction. Brookhaven Symp. Biol., No. 6, Abnormal and Pathological Plant Growth, 97—114.

— 1955: Resistance and susceptibility of carrot roots to crown-gall tumor formation. Proc. Nat. Acad. Sci. (U. S.) 41, 271—274.

— and J. L. Knupp, 1957: Sterile induction of crown-gall tumors on carrot tissues *in vitro*. Proc. Nat. Acad. Sci. (U. S.) 43, 199—203.

— and G. K. K. Link, 1952: Studies on the metabolism of plant neoplasms, V. Auxin as a promoting agent in the transformation of normal to crown-gall tumor cells. Proc. Nat. Acad. Sci. (U. S.) 38, 1066—1072.

— — 1955: The etiology of crown-gall. Quart. Rev. Biol. (Am.) 30, 207—277.

— E. M. Rasch, and H. Swift, 1953: Nucleic acids and tumor genesis in broad bean. Cancer Research 13, 499—502.

Kostoff, D., 1930: Tumors and other malformations on certain *Nicotiana* hybrids. Zbl. Bakter. usw., Abt. II, 81, 244—260.

— 1933: Tumor problem in the light of researches on plant tumors and galls and its relation to the problem of mutation. Protoplasma 20, 440—456.

KOSTOFF, D., 1939: Abnormal mitosis in tobacco plants forming hereditary tumours. Nature **144**, 599.
— 1943: Cytogenetics of the genus *Nicotiana*. States Printing House, Sofia. 1071 pp.
— and J. KENDALL, 1932: Origin of a tetraploid shoot from the region of a tumor on tomato. Science **76**, 144.
KRAUS, E. J., N. A. BROWN, and K. C. HAMNER 1936: Histological reactions of bean plants to indoleacetic acid. Bot. Gaz. **98**, 370—420.
KULESCHA, Z., 1947: Comparaison entre la structure anatomique des néoformations provoquées par l'action de l'acide indole-acétique et du *Phytomonas tumefaciens* sur des fragments de parenchyme vasculaire de Topinambour cultivés *in vitro*. C. r. Soc. Biol. **141**, 358—360.
— 1952: Recherches sur l'élaboration de substances de croissance par les tissus végétaux. Rev. Gén. Bot. **59**, 19—41, 92—111, 127—157, 195—208, 241—264.
— 1954: Croissance et teneur en auxine de divers tissus normaux et tumoraux. L'Année Biologique, 3e sér. **30**, 319—327.
— and R. GAUTHERET, 1948: Sur l'élaboration de substances de croissance par 3 types de cultures de tissus de Scorsonère: cultures normales, cultures de crown-gall et cultures accoutumées à l'hétéro-auxine. C. r. Acad. Sci. Par. **227**, 292—294.
KUNKEL, L. O., 1924: Studies on the mosaic of sugar cane. Bull. exper. Sta. Hawaiian Sugar Planters' Assoc., Bot. Ser., 3, part 2, 115—167.
— 1936: Heat treatments for the cure of yellows and other virus diseases of peach. Phytopathology **26**, 809—830.
— 1941: Heat cure of aster yellows in periwinkles. Amer. J. Bot. **28**, 761—769.
— 1954: Discussion. Brookhaven Symp. Biol., No. 6, Abnormal and Pathological Plant Growth, 127.
KUPILA, S., 1956: Anatomical and cytological development of crown gall. Arch. Soc. Zool. Bot. Fennicae "Vanamo" **10**, 38—50.
— 1958: Anatomical and cytological comparison of the development of crown gall in three host species. Ann. Bot. Soc. "Vanamo" **30**, 1—89.
KÜSTER, E., 1911: Die Gallen der Pflanzen. Leipzig.
— 1926: Regenerationserscheinungen an Bakteriengallen. Flora, N. F. **20**, 179—197.
LEE, A. E., 1952: Nitrogen and amino acids in normal, habituated, and bacteria-free crown gall tumor tissue cultures of grape. Plant Physiol. **27**, 173—178.
LEE, C. L., 1955 a: Virus-tumor development in relation to lateral-root and bacterial-nodule formation in *Melilotus alba*. Virology **1**, 152—164.
— 1955 b: Anatomical changes in sweet clover shoots infected with wound-tumor virus. Amer. J. Bot. **42**, 693—698.
— 1956: Virus-tumor formation in roots of *Capsella bursa-pastoris*. Phytopath. **46**, 140—144.
— and L. M. BLACK, 1955: Anatomical studies of *Trifolium incarnatum* infected by wound-tumor virus. Amer. J. Bot. **42**, 160—168.
LEVIN, I., and M. LEVINE, 1918: Malignancy of the crown gall and its analogy to animal cancer. Proc. Soc. exper. Biol. a. Med. (Am.) **16**, 21—22.
— — 1920: Malignancy of the crown-gall and its analogy to animal cancer. J. Canc. Res. (Am.) **5**, 243—260.
LEVINE, M., 1919: Studies on plant cancers. I. The mechanism of the formation of the leafy crown gall. Bull. Torrey Bot. Club **46**, 447—452.
— 1921: Studies on plant cancers. II. The behavior of crown gall on the rubber plant (*Ficus elastica*). Mycologia **13**, 1—11.
— 1923 a: Studies on plant cancers. IV. The effects of inoculating various quantities of different dilutions of *Bacterium tumefaciens* into the tobacco plant. Bull. Torrey Bot. Club **50**, 231—243.
— 1923 b: Studies on plant cancers. V. Leafy crown galls on tobacco plants resulting from *Bacterium tumefaciens* inoculations. Phytopath. **13**, 107—116.
— 1924: Crown gall on *Bryophyllum calycinum*. Bull. Torrey Bot. Club **51**, 449—456.
— 1925: A comparative cytological study of the neoplasms of animals and plants. J. Canc. Res. (Am.) **9**, 11—49.
— 1929: The chromosome number in crowngall and cancer tissues. Phytopath. **17**, 97.
— 1930: The chromosome-number in cancer tissue of man, of rodent, of bird and in crown gall tissue of plants. J. Canc. Res. (Am.) **14**, 400—425.
— 1931: Studies in the cytology of cancer. Amer. J. Cancer **15**, 144—211, 788—834, 1410—1494.

Levine, M., 1936: Plant tumors and their relation to cancer. Bot. Rev. **2**, 439—455.
— 1937: Tumors of tobacco hybrids. Amer. J. Bot. **24**, 250—256.
Limasset, P., and R. Gautheret, 1950: Sur le caractère tumoral des tissus de tabac ayant subi le phénomène d'accoutumance aux hétéro-auxines. C. r. Acad. Sci. Par. **230**, 2043—2045.
Link, G. K. K., and D. R. Goddard, 1951: Studies on the metabolism of plant neoplasms. I. Oxygen uptake of tomato crown-gall tissues. Bot. Gaz. **113**, 185—190.
Littau, V. C., and L. M. Black, 1952 a: Spherical inclusions in plant tumors caused by a virus. Amer. J. Bot. **39**, 87—95.
— — 1952 b: Spontaneous tumors in sweet clover. Amer. J. Bot. **39**, 191—194.
Locke, S. B., A. J. Riker, and B. M. Duggar, 1938: Growth substance and the development of crown gall. J. Agric. Res. **57**, 21—39.
— — — 1939: The nature of growth substance originating in crown gall tissue. J. Agric. Res. **59**, 535—539.
Loeb, L., 1937: The interaction between hereditary and stimulating factors in the origin of cancer. Acta Internat. Union Against Cancer **2**, 148—194.
— 1945: The biological basis of individuality. Charles Thomas, Springfield, Ill., and Baltimore, Md. 711 pp.
Lorz, A. P., 1947: Supernumerary chromonemal reproductions: polytene chromosomes, endomitosis, multiple chromosome complexes, polysomaty. Bot. Rev. **13**, 597—624.
Magnus, W., 1915: Durch Bakterien hervorgerufene Neubildungen an Pflanzen. S.ber. Ges. Naturfreunde zu Berlin, 263—277.
Magrou, J., 1927: Recherches anatomiques et bactériologiques sur le cancer des plantes. Ann. Inst. Pasteur **41**, 785—801.
Manigault, P., 1953: Étude biochimique et histochimique des tumeurs du crown-gall chez *Pelargonium zonale*. II. Phosphatases. Ann. Inst. Pasteur **85**, 602—620.
— A. Comandon et P. Slizewicz, 1956: Préparation d'un « principe inducteur » de la tumeur du *Pelargonium*. Ann. Inst. Pasteur **91**, 114—117.
Manil, P., 1950: Tumeurs végétales. Bull. Inst. Agron. et Stas. Recherches Gembloux **17** (1948—1949), 26—42.
Maramorosch, K., 1950: Influence of temperature on incubation and transmission of the wound-tumor virus. Phytopathology **40**, 1071—1093.
— M. K. Brakke, and L. M. Black, 1949: Mechanical transmission of a plant tumor virus to an insect vector. Science **110**, 162—163.
McEwen, D. M., 1952: Cancerous response in plants. Nature **169**, 839.
McKeen, W. E., 1954: An anatomical study of cane and crown galls. Canad. J. Bot. **32**, 527—530.
Miller, C. O., and F. Skoog, 1953: Chemical control of bud formation in tobacco stem segments. Amer. J. Bot. **40**, 768—773.
— F. Skoog, M. H. von Saltza, and F. M. Strong, 1955 a: Kinetin, a cell division factor from deoxyribonucleic acid. J. amer. Chem. Soc. **77**, 1392.
— — F. S. Okumura, M. H. von Saltza, and F. M. Strong, 1955 b: Structure and synthesis of kinetin. J. amer. Chem. Soc. **77**, 2662—2663.
Milovidov, P. F., 1930: Zur Zytologie der Pflanzentumoren. Protoplasma **10**, 294—296.
Morel, G., 1947: Transformations des cultures de tissus de vigne produites par l'hétéro-auxine. C. r. Soc. Biol. **141**, 280—282.
— 1948: Recherches sur la culture associée de parasites obligatoires et de tissus végétaux. Ann. Épiphyt. N. S. **14** (Sér. Path. vég. Mém. 5), 123—234.
Muir, W. H., A. C. Hildebrandt, and A. J. Riker, 1954: Plant tissue cultures produced from single isolated cells. Science **119**, 877—878.
Näf, U., 1958: Studies on tumor formation in *Nicotiana* hybrids. I. The classification of the parents into two etiologically significant groups. Growth (in press).
Naylor, J., G. Sander, and F. Skoog, 1954: Mitosis and cell enlargement without cell division in excised tobacco pith tissue. Physiologia Plantarum **7**, 25—29.
Newcomb, E. H., 1951: Effect of auxin on ascorbic oxidase activity in tobacco pith cells. Proc. Soc. exper. Biol. a. Med. (Am.) **76**, 504—509.
Nickell, L. G., 1952: Vitamin B_1 requirement of *Rumex* virus tumor tissue. Bull. Torrey Bot. Club **79**, 427—430.
— 1954: Nutritional aspects of virus-tumor growth. Brookhaven Symp. Biol., No. 6, Abnormal and Pathological Plant Growth, 174—186.
— 1955: Nutrition of pathological tissues caused by plant viruses. L'Année Biologique, 3e sér. **31**, 107—121.

NICKELL, L. G., P. GREENFIELD, and P. R. BURKHOLDER, 1950: Atypical growth of plants. III. Growth responses of virus tumors of *Rumex* to certain nucleic acid components and related compounds. Bot. Gaz. **112**, 42—52.

NITSCH, J. P., 1956: Methods for the investigation of natural auxins and growth inhibitors. p. 3—31, in: The Chemistry and Mode of Action of Plant Growth Substances. Proc. Symposium held at Wye College (Univ. of London) July 1955. Edited by R. L. Wain and F. Wightman.

NOËL, C., 1946: Recherches anatomiques sur le "crown gall." Ann. Sci. Nat., Bot., Sér. XI, **7**, 87—145.

PINOY, P. E., 1925: A propos du cancer des plantes ou crown gall. C. r. Acad. Sci., Par. **180**, 311—313.

PLATT, R. S., Jr., 1954: The inactivation of auxin in normal and tumorous tissues. L'Année Biologique, 3e sér. **30**, 349—359.

REINERT, J., and P. R. WHITE, 1956: The cultivation *in vitro* of tumor tissues and normal tissues of *Picea glauca*. Physiologia Plantarum **9**, 177—189.

RIKER, A. J., 1923 a: Some relations of the crowngall organism to its host tissue. J. Agric. Res. **25**, 119—132.

— 1923 b: Some morphological responses of the host tissue to the crowngall organism. J. Agric. Res. **26**, 425—436.

— 1926: Studies on the influence of some environmental factors on the development of crown gall. J. Agric. Res. **32**, 83—96.

— 1927: Cytological studies of crowngall tissue. Amer. J. Bot. **14**, 25—37.

— and T. O. BERGE, 1935: Atypical and pathological multiplication of cells approached through studies on crown gall. Amer. J. Cancer **25**, 310—357.

ROBINSON, W., 1927: Some features of crowngall in plants in reference to comparison with cancer. Proc. Soc. Med., Lond. (Sect. Trop. Dis. and Parasitol.) **20**, 1507—1510.

— and H. WALKDEN, 1923: A critical study of crown gall. Ann. Bot. **37**, 299—324.

ROSEN, H. R., 1926: Morphological notes together with some ultrafiltration experiments on the crown gall pathogene *Bacterium tumefaciens*. Mycologia **18**, 193—205.

SATINA, S., J. RAPPAPORT, and A. F. BLAKESLEE, 1950: Ovular tumors connected with incompatible crosses in *Datura*. Amer. J. Bot. **37**, 576—586.

SCHWARZ, K., R. DIERBERGER and A. A. BITANCOURT, 1955: Estudos sobre a cancer vegetal. I. A natureza quimica das auxinas de alguns tecidos vegetais normais e tumorais. Arq. Inst. Biol., São Paulo **22**, 93—118.

SEGRETAIN, G., 1949: Rôle de la plaie d'inoculation dans le développement des tumeurs à *Agrobacterium tumefaciens*. C. r. Acad. Sci. Par. **228**, 1452—1453.

SETÄLÄ, K., S. LUNDBOM und P. HOLSTI, 1957: Zur Anwendbarkeit von Schleimpilzen beim Studium des Mechanismus der Tumorgenese. Naturw. **44**, 285.

SHANTZ, E. M., and F. C. STEWARD, 1952: Coconut milk factor: the growth-promoting substances in coconut milk. J. amer. Chem. Soc. **74**, 6133—6135.

— — 1955: The identification of compound A from coconut milk as 1,3-diphenylurea. J. amer. Chem. Soc. **77**, 6351—6353.

SINNOTT, E. W., and R. BLOCH, 1945: The cytoplasmic basis of intercellular patterns in vascular differentiation. Amer. J. Bot. **32**, 151—156.

SKOOG, F., 1944: Growth and organ formation in tobacco tissue cultures. Amer. J. Bot. **31**, 19—24.

— 1954 a: Chemical regulation of growth in plants, p. 148—182, in: Dynamics of Growth Processes. Edited by E. J. Boell. Princeton Univ. Press, Princeton, N. J.

— 1954 b: Substances involved in normal growth and differentiation of plants. Brookhaven Symp. Biol., No. 6, Abnormal and Pathological Plant Growth, 1—21.

— and C. O. MILLER, 1957: Chemical regulation of growth and organ formation in plant tissues cultured *in vitro*. Soc. exper. Biol., Symp. XI, The Biological Action of Growth Substances 118—131.

— and C. TSUI, 1951: Growth substances and the formation of buds in plant tissues, p. 263—285, in: Plant Growth Substances. Edited by F. Skoog. Univ. Wisconsin Press, Madison, Wis.

SMITH, E. F., 1912 a: The staining of *Bact. tumefaciens* in tissue. Phytopathology **2**, 127—128.

— 1912 b: On some resemblances of crown-gall to human cancer. Science **35**, 161—172.

— 1913: Cancer in plants. Proc. 17th Internat. Congr. Med., Sect. III, Gen. Path. 281—298.

Smith, E. F., 1916 a: Further evidence that crown gall of plants is cancer. Science **43**, 871—889.
— 1916 b: Further evidence as to the relation between crown gall and cancer. Proc. Nat. Acad. Sci. (U. S.) **2**, 444—448.
— 1916 c: Studies on the crown gall of plants. Its relation to human cancer. J. Canc. Res. (Am.) **1**, 231—309.
— 1916 d: Crowngall studies showing changes in plant structures due to a changed stimulus. J. Agric. Res. **6**, 179—182.
— 1916 e: Crown gall and cancer. J. amer. med. Assoc. **67**, 1318.
— 1917: Embryomas in plants. (Produced by bacterial inoculations.) Bull. Johns Hopkins Hospital **28**, 277—294.
— 1920: An Introduction to Bacterial Diseases of Plants. W. B. Saunders Co., Phila., Pa., and London.
— 1921: Effect of crowngall inoculations on *Bryophyllum*. J. Agric. Res. **21**, 593—598.
— 1922: Appositional growth in crown-gall tumors and in cancers. J. Canc. Res. (Am.) **7**, 1—105.
— 1923: Twentieth century advances in cancer research. J. Radiol. **4**, 295—317.
— 1924: Crown-gall and its analogy to cancer. A reply. J. Canc. Res. (Am.) **8**, 234—239.
— 1925 a: Some newer aspects of cancer research. Science **61**, 595—601.
— 1925 b: Le cancer des plantes ou crown gall. Rev. Bot. Appl. et d'Agric. Col. **5**, 97—105.
— 1925 c: Cancer in plants and in man. Science **61**, 419—420.
— 1926: Recent cancer research. Amer. Naturalist **60**, 240—256.
— N. A. Brown, and L. McCulloch, 1912: The structure and development of crown gall: A plant cancer. U. S. Dept. Agric., Bur. Plant Indus. Bull. **255**, 61 pp.
— — and C. O. Townsend, 1911: Crown-gall of plants: Its cause and remedy. U. S. Dept. Agric., Bur. Plant Indus. Bull. **213**, 215 pp.
— and C O. Townsend, 1907 a: Ein Pflanzentumor bakteriellen Ursprungs. Cbl. Bakter. usw., Abt. II, **20**, 89—91.
— — 1907 b: A plant-tumor of bacterial origin. Science **25**, 671—673.
Smith, H. H., 1958: Genetic plant tumors in *Nicotiana*, in: Genetic Concept for the Origin of Cancer. New York Acad. Sci. (in press).
Snow, R., 1935: Activation of cambial growth pure hormones. New Phytologist **34**, 347—360.
Sonneborn, T. M., 1946: Experimental control of the concentration of cytoplasmic genetic factors in *Paramecium*. Cold Spring Harbor Symp. Quant. Biol. **11**, 236—255.
Sparrow, A. H., and J. E. Gunckel, 1956: Induction of tumours by ionising radiation on stems, leaves and roots of an interspecific *Nicotiana* hybrid, p. 485—488, in: Progress in Radiobiology. Oliver and Boyd, Edinburgh and London.
— — L. A. Schairer, and G. L. Hagen, 1956: Tumor formation and other morphogenetic responses in an amphidiploid tobacco hybrid exposed to chronic gamma irradiation. Amer. J. Bot. **43**, 377—388.
Spiegelman, S., 1954: Heritable differences in enzyme synthesizing capacity amongst cells of identical genotype. Proc. 2nd Nat. Cancer Conf., Ohio, Mar. 3—5, 1952, **2**, 1345—1349.
Stapp, C., 1938: Der Pflanzenkrebs und sein Erreger *Pseudomonas tumefaciens*. VI. *Asparagus sprengeri* Rgl. und *Phaseolus vulgaris* L. als Wirtspflanzen. Zbl. Bakter. usw., Abt. II, **99**, 116—123.
— 1956: Bakterielle Krankheiten. P. Sorauer, Handbuch der Pflanzenkrankheiten. 6. Aufl., Bd. 2, Lief. 2, Die Virus- und bakteriellen Krankheiten 567 pp.
Steward, F. C., and S. M. Caplin, 1951: A tissue culture from potato tuber: the synergistic action of 2, 4-D and of coconut milk. Science **113**, 518—520.
— — and E. M. Shantz, 1955: Investigations on the growth and metabolism of plant cells. V. Tumorous growth in relation to growth factors of the type found in coconut. Ann. Bot. N. S. **19**, 29—47.
— and E. M. Shantz, 1956: The chemical induction of growth in plant tissue cultures. p. 165—186, in The Chemistry and Mode of Action of Plant Growth Substances. Proc. Symposium held at Wye College (Univ. of London) July 1955. Edited by R. L. Wain and F. Wightman.
Stonier, T. T., 1955: Studies on crown gall. Thesis, Yale University.

STONIER, T. T., 1956 a: Labeling crown gall bacteria with P³² for radioautography. J. Bacter. (Am.) 72, 259—268.
— 1956 b: Radioautographic evidence for the intercellular location of crown gall bacteria. Amer. J. Bot. 43, 647—655.
STRUCKMEYER, B. E., A. C. HILDEBRANDT, and A. J. RIKER, 1949: Histological effects of growth-regulating substances on sunflower tissue of crown-gall origin grown in vitro. Amer. J. Bot. 36, 491—495.
SUIT, R. F., and E. A. EARDLEY, 1935: Secondary tumor formation on herbaceous hosts induced by Pseudomonas tumefaciens Sm. and Town. Scient. Agric. 15, 345—357.
SYLWESTER, E. P., and M. C. COUNTRYMAN, 1933: A comparative histological study of crowngall and wound callus on apple. Amer. J. Bot. 20, 328—340.
TANG, Y. W., and J. BONNER, 1947: The enzymatic inactivation of indoleacetic acid. I. Some characteristics of the enzyme contained in pea seedlings. Arch. Biochem. 13, 11—25.
TEITELBAUM, S. S., and L. M. BLACK, 1954: The effect of a phytophagous species of Tetrastichus, new to the United States, on sweet clover infected with wound-tumor virus. Phytopathology 44, 548—550.
THEIS, T. N., A. J. RIKER, and O. N. ALLEN, 1950: The destruction of crown-gall bacteria in periwinkle by high temperature with high humidity. Amer. J. Bot. 37, 792—801.
THERMAN, E., 1951: The effect of indole-3-acetic acid on resting plant nuclei. I. Allium cepa. Ann. Acad. Sci. Fennicae, Ser. A, IV. Biol. 16, 1—40.
— 1956: Dedifferentiation and differentiation of cells in crown gall of Vicia faba. Caryologia 8, 325—348.
THOMAS, J. E., and A. J. RIKER, 1948: The effects of representative plant growth substances upon attenuated bacterial crown galls. Phytopathology 38, 26.
THOMAS, P. T., H. J. EVANS, and D. T. HUGHES, 1956: Chemically induced neoplasms in fungi. Nature 178, 949—951.
WHITAKER, T. W., 1934: The occurrence of tumors on certain Nicotiana hybrids. J. Arnold Arboretum 15, 144—153.
WHITE, P. R., 1939: Controlled differentiation in a plant tissue culture. Bull. Torrey Bot. Club 66, 507—513.
— 1944: Transplantation of plant tumors of genetic origin. Cancer Research 4, 791—794.
— 1945: Metastatic (graft) tumors of bacteria-free crown-galls on Vinca rosea. Amer. J. Bot. 32, 237—241.
— 1951: Neoplastic growth in plants. Quart. Rev. Biol. 26, 1—16.
— 1954: The Cultivation of Animal and Plant Cells. Ronald Press Co., New York. 239 pp.
— and A. C. BRAUN, 1942: A cancerous neoplasm of plants. Autonomous bacteria-free crown-gall tissue. Cancer Research 2, 597—617.
— and W. F. MILLINGTON, 1954 a: The distribution and possible importance of a woody tumor on trees of the white spruce, Picea glauca. Cancer Research 14, 128—134.
— — 1954 b: The structure and development of a woody tumor affecting Picea glauca. Amer. J. Bot. 41, 353—361.
WINGE, Ö., 1927: Zytologische Untersuchungen über die Natur maligner Tumoren. I. "Crown gall" der Zuckerrübe. Z. Zellforsch. usw. 6, 397—423.

.